Intelligent Systems Reference Library

Volume 169

Series Editors

The aim of this series is to publish a Reference Library, including novel advances and developments in all aspects of Intelligent Systems in an easily accessible and well structured form. The series includes reference works, handbooks, compendia, textbooks, well-structured monographs, dictionaries, and encyclopedias. It contains well integrated knowledge and current information in the field of Intelligent Systems. The series covers the theory, applications, and design methods of Intelligent Systems. Virtually all disciplines such as engineering, computer science, avionics, business, e-commerce, environment, healthcare, physics and life science are included. The list of topics spans all the areas of modern intelligent systems such as: Ambient intelligence, Computational intelligence, Social intelligence, Computational neuroscience, Artificial life, Virtual society, Cognitive systems, DNA and immunity-based systems, e-Learning and teaching, Human-centred computing and Machine ethics, Intelligent control, Intelligent data analysis, Knowledge-based paradigms, Knowledge management, Intelligent agents, Intelligent decision making, Intelligent network security, Interactive entertainment, Learning paradigms, Recommender systems, Robotics and Mechatronics including human-machine teaming, Self-organizing and adaptive systems, Soft computing including Neural systems, Fuzzy systems, Evolutionary computing and the Fusion of these paradigms, Perception and Vision, Web intelligence and Multimedia.

** Indexing: The books of this series are submitted to ISI Web of Science, SCOPUS, DBLP and Springerlink.

More information about this series at http://www.springer.com/series/8578

Parikshit N. Mahalle · Prashant S. Dhotre

Context-Aware Pervasive Systems and Applications

Springer

Parikshit N. Mahalle
Computer Engineering
Shrimati Kashibai Navale
College of Engineering
Pune, Maharashtra, India

Prashant S. Dhotre
Computer Engineering
Dr. D. Y. Patil Institute of Technology
Pimpri, Pune, Maharashtra, India

ISSN 1868-4394 ISSN 1868-4408 (electronic)
Intelligent Systems Reference Library
ISBN 978-981-32-9954-2 ISBN 978-981-32-9952-8 (eBook)
https://doi.org/10.1007/978-981-32-9952-8

This Springer imprint is published by the registered company Springer Nature Singapore Pte Ltd.
The registered company address is: 152 Beach Road, #21-01/04 Gateway East, Singapore 189721, Singapore

Dedicated to

My Beloved Parents
Late Dr. Narendra L. Mahalle
Smt. Rajani Narendra Mahalle
My Wife Namita, Daughter Yashita
&
The Almighty God

Parikshit N. Mahalle

My Beloved Parents
Dr. Shantaram M. Dhotre
Mrs. Shashikala S. Dhotre
My Wife Deepali, Beloved sons Chaitanya and Harsht
&
The Almighty God

Prashant S. Dhotre

Preface

Karma

The meaning of Karma is in the intention. The intention behind the action is what matters. Those who are motivated only by the desire for the fruits of action are miserable, for they are constantly anxious about the results of what they do.

Bhagwad Gita

This is envisioned to be a simple, accessible book and key reference which puts forwards best research road map, challenges and future outlook on context-aware pervasive systems and applications. Due to broadband technology and wide availability of the Internet, Internet computing has been evolved into the Internet of things and eventually into an Internet of everything. Modern scientific, engineering and business applications are increasingly dependent on a machine-to-machine communication wherein there is less human intervention. This leads to the need for context-aware pervasive systems and applications where RFID, sensors and smartphones will play a key role to sense the context. The book focuses on an overview of context and context management and how to perform context management in different use cases. In addition to this, the book also covers context-aware computing and personalization, various architectures for context-aware systems and security issues. This book emphasizes on the overview of context-aware pervasive systems, several case studies to visualize the use of the pervasive system, challenges and the future aspects of context-aware pervasive systems and applications.

The main characteristics of this book are:

- It assumes that the reader's goal is to achieve a complete understanding of context management issues, challenges and possible solutions in pervasive systems. It is not any specific use case oriented, but also context-aware computing problems are discussed across various domains.

- This book is motivating to use context management for smart computing paradigm, in new inventions for a wide range of stakeholders like a layman to educate users, villages to metros and national to global levels.
- This book contains numerous examples, case studies, technical descriptions, scenarios, procedures, algorithms and protocols. These deliverables have been developed with utmost care, and they will be useful to the readers in a broad range of applications in context-aware computing.
- The main endeavour of this book is threat analysis and activity modelling of attacks using the use case approach in order to give an actual view of the context-aware pervasive systems. This unique approach will certainly help readers for better understanding.

Chapter 1 gives an understanding of pervasive systems, emerging trends and economic significance. The role of sensors, actuators, RFID and Internet of things as technical building blocks in context-aware systems is also introduced in this chapter. Several issues and challenges in the context of design, technology and security are also discussed in this chapter. Finally, Chap. 1 concludes with an overview of the entire book.

Chapter 2 presents an overview of context, its classification and the notion of context-aware computing. In this chapter, the state of the art in sensing the context using RFID, sensors and ontology is also presented and discussed. Chapter 2 finally concludes with a requirement for representing and exchanging the contexts in pervasive systems.

Chapter 3 describes the basic elements of context-aware pervasive systems in general. A discussion on context-aware networks along with the state of the art is also presented in the next part of this chapter. The issues and challenges with respect to routing, mobility and data management are also discussed in the last part of Chap. 3.

Chapter 4 explains how context-aware computing is important for personalization. The relationship between machine learning, ontology learning and personalization is also discussed in this chapter. The next part of this chapter focuses on context-aware profiling and profile translation towards context-aware services. The need for clustering in ubiquitous computing and how context-aware clustering will play an important role in the next-generation Internet is discussed in the last part of this chapter.

Chapter 5 initially presents architectural requirements, different approaches and various platforms for context-aware applications. This chapter also presents various context-aware algorithms along with design process and system issues. The last part of this chapter presents various case studies of context-aware systems and machine learning.

Chapter 6 describes telecommunication context-aware services like content, session, location and identity. Telecom service-oriented architecture is presented and discussed in the next part of this chapter. The context-aware business model is also discussed in this chapter.

Chapter 7 discusses various security issues in context-aware systems. The importance of trust and security in pervasive systems is also presented in this chapter. The next part of this chapter focuses on very important issues of activity modelling and threat taxonomy in context-aware systems as well as corresponding mitigations.

Finally, Chap. 8 summarizes the book and concludes by proposing future work, which can be researched and build based on the ideas and challenges presented. Several applications of context-aware pervasive systems are also presented as concluding remarks in this chapter.

Although context-aware systems are primarily for undergraduate students, the book is accessible to a wider audience of researchers and designers who are interested in exploring pervasive computing further. The book is useful for undergraduates, postgraduates, industry, researchers and research scholars in ICT, and we are sure that this book will be well-received by all stakeholders.

Pune, India
November 2019

Parikshit N. Mahalle
Prashant S. Dhotre

Acknowledgements

I would like to acknowledge and appreciate the support of my co-author Dr. Prashant S. Dhotre for his contribution towards improving the value of this book. I am very much thankful to Prashant for being proactive and instrumental in providing his thoughts and suggestions to address the best practices in this book, wonderful coordination and generous response. I am proud of Prashant for being a technical reviewer of this book.

I would like to acknowledge and thank my Ph.D. supervisor, Associate Professor Nile Rashmi Prasad, immensely for her tireless and unconditional help, support, guidance and being a role model for me. I am highly indebted to my colleagues and friends Dr. Gitanjali R. Shinde, Dr. Vinod.V. Kimnahune and Dr. Shafi.K. Pathan for their encouragement and inspirational support. I am also thankful to my Ph.D. student Poonam N. Railkar, master's students Sandesh Mahamure and Ganesh Pise for their help and support in drafting contents and drawing figures.

I would like to thank my mother for supporting my academic endeavours. I am thankful to her for being a backbone to me always. I am deeply indebted to my father Late Dr. Narendra Mahalle for his inspiration and love. He would have been very happy today. Lastly but most importantly, I want to thank my loving wife Dr. Namita. She has tirelessly supported me while writing this book. I would also like to thank my daughter Yashita for making me forget all the pressure with her innocent smile. I would also like to thank all those who directly and indirectly involved in building this book and research work.

Parikshit N. Mahalle

I am extremely grateful to my guru and co-author Dr. Parikshit N. Mahalle for his valuable support, scholarly inputs, motivation and constant encouragement for completion of this book. Dr. Parikshit N. Mahalle has been a very passionate and positive persona. I consider it is an amazing opportunity to do write a book with his directions and to learn from his research expertise. He has been a technical reviewer that helped me to sharpen the contents of this book. I am thankful to him for his all comments and compliments.

I must thank my Ph.D. supervisor, Associate Professor Dr. Henning Olesen, who keeps me involved in the research work. His discipline, perfections, tireless and unconditional support helps me to continue research work. I take this opportunity to thank my colleague and friend Prof. Mahendra B. Salunke for his encouragement and inspirational support.

I would like to thank my father for supporting my academic endeavours. I always learn positiveness towards life from him. I always admire my mother for being patience in every situation. Lastly but most importantly, I want to thank my ambitions and loving wife Prof. Deepali. She has tirelessly supported me while writing up this book. I would also like to thank my two lovely sons Chaitanya and Harsheet for making me forget all the pressure with their innocent smile. I would also like to thank all those who directly and indirectly involved in building this book and research work.

Prashant S. Dhotre

We could not have written this book without the encouragement of our well-wishers and family members. We would like to thank the many people who encouraged and helped us in various ways throughout this book, namely our colleagues, friends and students. We are indebted to Honourable Founder President of STES, Prof. M. N. Navale, Founder Secretary of STES, Dr. Mrs. S. M. Navale, Vice President (HR), Mr. Rohit M. Navale, Vice President (Admin), Ms. Rachana M. Navale, our Principal, Dr. A. V. Deshpande, Vice Principal, Dr. K. R. Borole, Dr. K. N. Honwadkar, for their constant encouragement, inexplicable support and faith on us.

We would like to thank the Honourable President of Dr. D. Y. Patil Unitech Society, Dr. P. D. Patil, for continuous support and guidelines. I am extremely thankful to Dr. Suresh N.Mali, Principal of Dr. D. Y. Patil Institute of Technology (DIT), Pimpri, Pune, for his praising words and inexplicable support. I also would like to thank Dr. Pramod D. Patil (Vice Principal) for the kind support.

We are also very much thankful to all our department colleagues at SKNCOE and DIT for their continued support, help and keeping us smiling all the time.

Last but not least, our acknowledgements would remain incomplete if we do not thank the team of Springer Nature who supported us throughout the development of this book. It has been a pleasure to work with Springer Nature team, and we extend our special thanks to the entire team involved in the publication of this book.

Finally, we would like to thank Mr. Aninda Bose, Mr. Ramamoorthy Rajangam and Ms. Shilpa of Springer Nature for their continuous support in completing this book.

Our special thanks to Dr. Nilanjan Dey for his continuous support and guidance always.

Parikshit N. Mahalle
Prashant S. Dhotre

Contents

About the Authors

Dr. Parikshit N. Mahalle obtained his BE degree in CSE from Sant Gadge Baba Amravati University, India, and his ME degree (in Computer Engineering) from Savitribai Phule Pune University (SPPU), Pune, India. He completed his Ph.D. at Aalborg University, Denmark. Currently, he is working as Professor and Head of the Department of Computer Engineering at STES Smt. Kashibai Navale College of Engineering, Pune, India. He has over 18 years of teaching and research experience.

He is a senior member of the IEEE, member of the ACM, and life member of the CSI and ISTE. He serves as a reviewer, Associate Editor and member of the Editorial Review Board for various journals. Dr. Mahalle has 140+ research articles and eight books to his credit, and has edited three books by Springer and CRC Press. He received the "Best Faculty Award" from STES and Cognizant Technologies Solutions.

Dr. Prashant S. Dhotre is an Associate Professor at the Department of Computer Engineering at Dr. D. Y. Patil Institute of Technology, India. He obtained his BE degree in CSE from SRTMU, Nanded, and his ME degree in IT from the SPPU, Pune. He subsequently completed his Ph.D. degree at Aalborg University, Denmark.

He has over 15 years of teaching and research experience, and has spearheaded research projects on "privacy and security" funded by the SPPU, Pune. He has more than 30 articles and three books to his credit, and received the "Best Paper" award in an ICICC—2017 Springer Series. He also serves as a reviewer for the various journals.

Dr. Dhotre is currently serving as SPOC for Foreign Collaboration between Indian Institutes (Sinhgad, Dr. DY Patil Pimpri, MIT ADT, etc.) and AAU, CPH, Denmark [Outcome: 70+ Indian students completed internships and two candidates completed their postdocs].

List of Figures

List of Tables

Chapter 1
Pervasive System Overview

Abstract Communication has been improved, and services have been enriched with the help of revolutionary smart objects and computing environment that assist the users to make daily tasks much easier. The vision of pervasive system imposes on the creating of an environment that has communication and computing capabilities with seamless integration of human interventions. Looking at the users' requirements and required services during several scenarios, the pervasive system is an essential and effective solution that needs to discuss. Personalization and context-based services are the need of an hour. This chapter starts with the vision of pervasive system and many emerging trends. The chapter continues to discuss how the pervasive systems not only benefit the users but also businesses. The next part of this chapter focuses on the basic technical building blocks. The issues and challenges of the pervasive system are presented in this chapter. The overall overview of this book is discussed in the last part of this chapter.

1.1 Overview

This section describes the overall view of the pervasive system, the technical side of the system and issues and challenges of the pervasive system.

1.1.1 Pervasive System—Vision

Computers and mobile phones have become the essential components of our life. Using wireless and mobile communication technologies, user's life has become convenient and efficient. The online services that have brought up to the users include booking a movie ticket, reading a book, sharing photographs with relatives, chatting with friends and playing games. However, this has just touched up the outer surface of the intelligent world formed by the integration of computers and mobile phones.

These small computers and mobile phones with the help of existing IT environments are becoming revolutionary and can change the businesses, life of people,

P. N. Mahalle and P. S. Dhotre, *Context-Aware Pervasive Systems and Applications*,
Intelligent Systems Reference Library 169,
https://doi.org/10.1007/978-981-32-9952-8_1

etc. With the help of embedding computational device into the existing objects, the revolutionary objects are becoming effective in communication with other devices and able to perform useful tasks that minimize user interaction with computers.

This was well envisioned by the Mark Weiser in 1991. He stated the vision of pervasive computing (earlier ubiquitous computing) as [1]:

> The most profound technologies are those that disappear. They weave themselves into the fabric of everyday life until they are indistinguishable from it.

The core of above-stated vision emphasizes on the formation of an environment that has communication and computing capability with equal prospects of human integration. This vision was stated well prior to the existence of the required hardware. After an era of hardware and technology progress, the required elements of pervasive systems were available. The elements or the products could be wearable, and handheld computers, wireless LANs and intelligent devices have sensing and control capabilities. This implies that we are well positioned in the era where we could start to pursue the vision stated by the Weiser [1].

Contrasting with a computer or laptop computing, the pervasive system can be implemented in any device, in any place, at any time. The devices can communicate any data with other devices or systems across any network.

The evolution of devices from a computer, laptop or notebooks to wearable devices, RFID, sensors, etc., is contributing in making the pervasive system effective. The main goal of the pervasive system is to make 'smart' devices that enable to create an environment capable of sending, collecting, storing, processing and communicating as a way to understand the context of collected data and concern activity. Thus, the pervasive systems understand the context (surrounding) and enhance user experiences and life.

The application domains of the pervasive system include health care, military, consumer and logistics. The known example is Apple Watch [2] which a smart and healthy watch. This watch informs the users about phone calls and allows the call to complete through it. Also, this watch keeps a track of your runs, monitors health rate, connects to your car, transfers money, etc.

There are several universities and organizations that have emerging projects of pervasive systems. This includes 'Aura' from Carnegie Mellon [3], 'Oxygen' from MIT [4] and 'Endeavor' by UC Berkeley [5]. The projects from the above stated have addressed different issues of pervasive computing systems. The efforts put in by each of the projects had intent to hit the pervasive system a reality.

To exist in a world with pervasive computing, we are presenting two scenarios that will give you a look as well as a feel. The scenarios are selected in such a way that it will happen in the coming years. We introduce 'Swayam' as a pervasive system in the following scenarios.

Scenario 1

> *Mr Kiran is a businessman travelling to the US and he is at Frankfurt airport to catch up his connecting flight. During the layover, at a Gate 25 in the airport, he has modified his*

business proposal and wishes to send it to the client using local wireless communication. Unfortunately, the wireless connection is weak at that place. The 'Swayam' observes that Mr Kiran is unable to share the proposal.

The 'Swayam' communicates with the airport's network and identifies Mr Kiran's next journey and finds the location (coffee shop at the airport) where the signal strength is strong. Immediately, 'Swayam' runs a dialogue box on Mr Kinan's laptop suggesting that Mr Kiran can go to the nearest coffee shop which just 5 min distance by walk and also inform that a special discount offered if he buys a coffee.

Gracefully, Mr Kiran accepts the suggestion and complete his task at a coffee shop while enjoying the coffee. The 'Swayam' send a task confirmation message as an acknowledgement. The 'Swayam' informs Mr Kiran the connecting flight is on time, therefore, Mr Kiran comes to the Gate 25 for his ongoing journey.

Scenario 2

Ms Anjali is in her researchers' room, preparing for her PhD lecture at which she will present her research work to a committee. The PhD lecture room is in another building on the same campus. The PhD lecture should start within a few minutes, but Ms Anjali is not ready. She picks her handheld mini-computer and walks out towards the presentation room. The 'Swayam' immediately transfer the presentation from her desktop computer to her handheld minicomputer and allows her to make the updates/changes in her presentation using her voice commands.

Looking at Ms Anjali's calendar, the 'Swayam' concludes where Ms Anjali is going. The 'Swayam' identifies the projection computer is not connected to the projector. Immediately, the technical team was called by 'Swayam' before Ms Anjali reaches presentation location.

As Ms Anjali reaches the presentation location, the 'Swayam' downloads the presentation from her minicomputer to the projection computer. Ms Anjali observes the presentation with all edits is updated. Ms Anjali delivers her PhD lecture on a high note, leaving the committee members impressed by her research work and polished presentation.

Learning from Scenarios

Above scenarios highlight several key concepts in the pervasive system/computing. The first concept observed from Scenario 1 is 'proactiveness'. Mr Kiran has been able to complete the e-mail transmission since the 'Swayam' proactively identified the problem which could delay the task by several minutes. The 'Swayam' is also able to combine several pieces of knowledge together to guide/help Mr Kiran. The knowledge includes detecting weak Internet connectivity, finding the nearby location(s) where Internet connectivity is strong and what Mr Kiran like so that it can be offered to him.

Scenario 1 also depicts the importance of understanding the context (time, preferences and location). The 'Swayam' can know the information about the flights, the gates, the time between connecting flights, etc., at the same airport.

Scenario 2 represents the cross-platform functionality with fewer efforts. The self-tuning is the ability of pervasive computing. Ms Anjali could able to complete the task as the 'Swayam' took her presentation from her desktop computer to a minicomputer. Automatic adjustment as per the requirements is very well represented in the scenario where the voice commands are read and converted in the required format. In this scenario also, the proactiveness is shown by 'Swayam' at several instances of the time like identifying the schedule of Ms Anjali, communicating with the projector, computer, calling technician well in advance, transferring the revised presentation to projector computer.

Scenario 2 also depicts the importance of the context (location, preferences and time). The 'Swayam' knows the presentation schedule, the issue with the projector computer, etc., on the same campus.

1.1.2 Emerging Trends

In the mid-1970s, the evolutionary step in computing was pervasive computing. Before pervasive computing, the important two distinct steps are computer networks, distributed systems and mobile computing. Before we start discussing the pervasive computing and solution it provides, we must understand each evolution represented in Fig. 1.1 and the intellectual relationship among these steps.

Computer Network

An interconnected collection of autonomous computers that can exchange data is called computer networks. Two or more computers are connected to each other for doing a certain task is achieved using computer networking. In personal computers (PCs), a personal task like word processing would not require computer networking. However, information and communication are important concerns in any task. The businesses can get benefits if your computer is able to communicate with other computers/systems. The communication could be sharing files, specific media, etc.

In the early days of the computer era, computers communicate with another computer using wires. The data to send from a computer was lined up in the message queue and sends it to another computer as fast as possible using wired media.

The local area network (LAN) is the communicating computers which may be in the same building or in the same area or in the same city. If the computers are in the same building, then they are connected (wired) that run across the same building. This could be achieved with the help of a leased line provided by the telephone company. This could be expensive as it requires to be 'always on'.

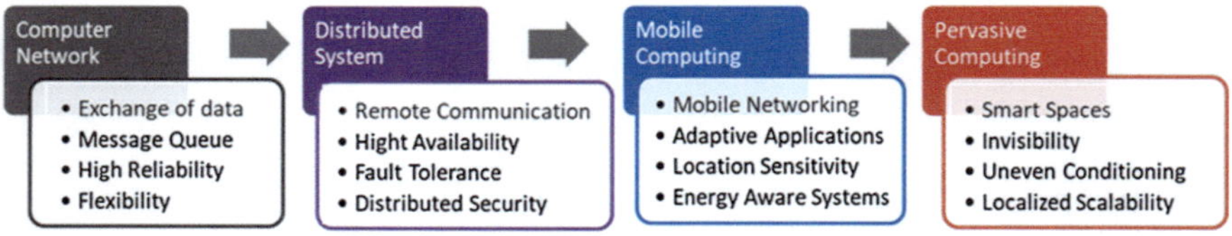

Fig. 1.1 Evolution of pervasive computing/system [4]

If the computers who wish to communicate are in far away, the leased line is extended to a certain extent and controlled by central offices. The direct connection between two computers situated in different cities could be very expensive. If your computer (A) is connected to another computer (B) and that computer (B) is connected to another computer (C) and so on, you can send the data as long as such computers (A, B, C, etc.) are connected to each other. Also, these computers must agree to store and forward the data (message). Hence, using the shared software and hardware resources, the cost is reduced. This provides advantages like cost reduction, high reliability, flexibility, etc.

However, if a user who wishes to communicate with another user, then he/she has to login onto one computer. Once logged in, the same user needs to do the job explicitly. The users should explicitly move the files around the computers.

Distributed System

The PC and LAN continued to extend to match up the requirements from 1970 to early 1990. The requirement leads to the combination of personal computers and computer networks (LAN, MAN, etc.). This combination had given rise to the distributed systems. The distributed system is a conceptual framework that has provided a great value where the static or mobile devices are connected to each other. Also, the media could be wired or wireless communication, and information sharing is vital.

In a computer network, the computers are connected to each other physically. There is no middleware between the computers that process and manage the communication. The specific requirement that has given rise the distributed systems is that the transparency in the communication. A distributed system allows computers to communicate and able to accomplish an aim. An example could be running a database on one computer, the office software that makes use of the database running on another computer and the user who is making updates in the data is working from the third computer.

It is an interconnection of autonomous computers through a middleware is a distributed system.

The difference between the computer network and distributed systems is the one who invokes the movement of the files/data. Is it a user or operating system? In computer networks, the user is responsible for the movement of data from one computer to another computer. However, on the behalf of the user, the operating system/software moves the files in a distributed system. The knowledge span of a distributed system is wider, and it has been the foundation to pervasive system/computing, and it is represented in Table 1.1.

Mobile Computing

In the early 1990s, the rapid development of handheld devices had given birth to the mobile computing field. Looking at the functionalities provided by the distributed systems with laptop computers, the challenge was to see the working of the same functionalities for mobile clients. Even the basic principles of the distributed system have been applied to the mobile computing field. However, due to the dynamics of mobile devices and computing power, there were several issues resulted in the

Table 1.1 Knowledge span of distributed system

Sr. No.	Features	Explanations
1	Remote communication and access	These include communication to a remote computer and access the contents from the database. This could be achieved with the help of a remote procedure call and protocol layering [6]. In the distributed system, the database must support the distributed databases [7]
2	Fault tolerance	This includes the value to the transactions by confirming its complete occurrence or complete failure of the transaction. The transaction can be distributed or nested transactions that need to be considered carefully in the distributed system. Also, it is important to consider the integrity of the data in the distributed system. Integrity is achieved with the help of two-phase commit [8]
3	High availability	This includes the important aspect of 'always on' in the distributed system. The system should be available with the help of replica control [9] and mirrored execution [10]
4	Security	This includes the standard measures that provide security to the distributed systems. The security measure should include the encryption-based authentication protocol [11]. The privacy protection is also an important issue that should adhere to the privacy principles as mentioned by Privacy by Design (PbD) [12]

distributed system with mobile devices/clients. The issues include appearances of the functionality provided by the distributed system.

There were several constraints like the volatile nature of network quality, lowered robustness of the mobile devices, limitation due to mobile size and weight, battery power consumption, etc. [13]. Such constraints have forced the researchers to come up with specialized technique development. The knowledge span of mobile computing is much wider, and it is represented in Table 1.2.

Pervasive Computing

At the beginning of this chapter, we already described pervasive computing as an environment that has communication and computing capability. This is elegantly integrated with the users so that it simply becomes 'technology that disappears'. Even mobile computing—a previous step in the latest trends of technology—should support mobility. As per the user's life is considered, the motion is a vital part of his/her life. Hence, the technology should support the mobility; otherwise, the

Table 1.2 Knowledge span of mobile computing

Sr. No.	Features	Explanations
1	Mobile networking	This includes the addressing of mobile clients. Since the mobile devices have several constraints like hardware, operating system, vendor, mobility, etc., there is a need to understand the addressing mechanism in the distributed environment. Generally, it is well supported with the help of ad hoc protocols [14] and mobile IP [15]. To improve the TCP performance in wireless environment, few techniques are proposed and implemented [16]
2	Support for adaptive applications	Looking at the mobility in mobile computing, the applications need to be executed without any fault. The mechanisms need to be fast enough to detect and adopt the mobility like application-based agile adaptation in mobile computing [17] and transcoding by proxies [12]
3	Energy-saving techniques	This includes the techniques that consider the issues or constraints of mobile or tiny devices in mobile computing. The energy or power is the main concern in the mobile environment. Hence, the mechanisms adapt the application and saves energy [18] The mobile devices also need proper scheduling with time-efficient mechanisms. Hence, this is the need of the twenty-first century where the mobile task needs to schedule so that the processor needs to make a decision by varying speed of processor scheduling [19] Memory is another constraint in mobile computing where the effective space management of those mobile devices. The mechanism also should be energy efficient [20]
4	Location sensitivity	This includes the important context called location. The location is also an important value in pervasive computing. Sensing the location of mobile devices in mobile computing is vital to understand the requirement to process an application. The application which is based on location called location-based spatial queries (LBSQs) [21]. The techniques are promising [22]. The nature of the mobile device is important to understand the behaviour of the devices which is based on location. Hence, the behaviour of the system is identified based on its location [23, 24]

technology without motion support is like a user is aware of the technology but cannot use it when a user moves.

Therefore, the research agenda of pervasive computing is in line with mobile computing but moves further. The extensive research includes following thrusts as a part of the research agenda that may not have completely addressed in mobile computing. The knowledge span of the pervasive system is represented in Table 1.3.

Table 1.3 Knowledge span of pervasive computing

Sr. No.	Features	Explanations
1	Effective use of smart spaces	This is the first point in the research agenda of pervasive computing. There are several ways to define space in pervasive computing. This could be a bounded area like a passage, or a classroom or meeting room, or an open area such as a plot A smart space is integrating building infrastructure with computing or intelligent infrastructure. Surprisingly, these two infrastructures were separate until now [25]. This has given an opportunity to both the environments to control and sense each other There are several examples of smart spaces where controlling and sensing are performed. A simpler example is controlling home appliances or lighting system based on the electronic profile of the homeowner. Hence, the user's location is crucial to decide the operations or decide the preferences. On the other side, the behaviour of the software or application depends on the user and the user location. The user and other objects in the smart space define the smartness
2	Invisibility	The second point in the research agenda of pervasive computing is invisibility. In the ideal case, the technology that disappears is expressed by Weiser. The invisibility is expressed from the user's perspective or the user's consciousness. In the reel case, the less user interaction with the technology which is the approximation to the ideal case. In practice, a reasonable approximation to this ideal is minimal user distraction Over the period, if the pervasive computing satisfies the user needs and in a few cases, the user needs interaction with technology, then the technology has achieved the subconscious level of a user [26]

(continued)

Table 1.3 (continued)

Sr. No.	Features	Explanations
3	Localized scalability	The third research agenda point in pervasive computing is localized scalability. As the users' smart space grows at a high level, then personal computing space and users surrounding interaction increase at a high-intensity level. This could lead to several issues like energy, bandwidth and distance management. Even these issues become critical as the number of users increases Hence, at the outset, the scalability is a critical issue in pervasive computing. The literature has tried to address this issue but still, the physical distance needs to consider at a deeper level. For example, the server should be able to process the user request despite the location of the user (near or very far). As the user moves away, the density of interaction should also minimum with the home environment
4	Masking variance in the conditioning	This is forth research agenda in the pervasive computing the variance and level of masking it. The computability or the use of pervasive computing may vary as per the infrastructure available. This infrastructure variance depends on the business models, economics. Hence, the level of smartness varies from the well available non-technical factors. This could be a staff room, office or conference room as compared to the other locations Designing personal computing or pervasive computing for the worst situation may reduce the level of variance on the condition of the available environment. Remember, the complete invisibility is not possible, but limiting the variability to well level is possible

1.1.3 Economic Significance

Pervasive computing provides an environment that senses the presence of the user/individual, detects and accepts the requirements, and intelligently responds to the user. The human-centred technology is depending on the intelligent system that provides useful information to the user in such a way that the technology looks invisible to the user. Not only a user, the pervasive computing and its capabilities can benefit the businesses. Though it is a key challenge, however, the main aspect like pervasive communication, use of sensors and actuators and embedding intelligence in the infrastructure can achieve the vision.

The extended ability of pervasive computing can enable communication and interaction between anything (or anyone) who are connected to the pervasive system. Each thing or entity who has the capability of communication and computing ability can initiate a communication, accept it, interpret it and subsequently process it. This leading vision creates value and helps in organizing businesses.

Create new values or new businesses are possible with the help of new infrastructure development or creating new applications that support both the organizations as well as users. There are several business domains where pervasive computing can move the economics like communication and messaging, entertainment, e-governance, safety, health care, information services, education, monitoring/quality of life, banking, etc.

Some of the basic applications can be explained as:

- *'Digital Reply'*
  If an application *'Digital Reply'* on your cell phone detects incoming calls, then it responds to it using your voice commands. Also, this application can give an appropriate reply to the text received on the user's cell phone.
- *'Safe Home'*
  A device 'Safe Home' mounted on a wall outside the main door of your home. An unknown (unacquainted) person is at your doorstep and he is pressing a doorbell. As this person is new to the 'device', it does not identify him/her. Therefore, a text message is created by the device mentioning a new person is knocking the door. The owner will receive the information about the new person at his/her door. This could help the homeowner for verification and additional information. This could provide home security.
- *'Make me Happy'*
  A device/application 'Make me Happy' come to know that you are coming home from the office in the evening with your other information like your physical state (heart rate, blood pressure, etc.) as well as mental state (happy, sad, etc.). According to this collected information, the application alters the lighting system at home, start brewing the coffee and turn on/off the electric geezer. Also, the news and TV channels are set according to the mood.
- *'I'll take care'*
  There are several instances where you need extra care. There are also some situations where you are going for a business tour and you forget to collect the necessary stuff. The application 'I'll take care' will notify you about those stuff before you leave your home/office to business tour.
- *'Hello Doctor'*
  Imagine you have a meeting, but you are ill, the application 'Hello Doctor'—on your behalf—will communicate to the meeting coordinator that you are not well and perhaps you will not attend a meeting. The same application will collect your health information and will share with your family doctor.

- *'My Agent'*
  For personal or family work sometimes, you need a personal device/application 'My Agent' that helps you for ordering the groceries on your behalf. This 'agent' could also identify that your health insurance is expired. It also looks for a better and cheaper insurance policy for you.

Looking at the various domains and the applications mentioned above, pervasive computing is appealing, and it has the ability to give great value to the businesses. All the above applications use local communication (within a specified area) and the intelligence embedded in the device. However, the 'context awareness' is the crucial aspect of developing such devices/applications. The applications are proactive or reactive, depending on the state of the user.

The wide range of innovative and effective applications reflects greater business opportunities. The new business ideas will try to destroy the stability in the businesses. This innovation will keep the businesses to run over a long time, and the companies can set long-term goals.

The economics will keep on changing as the user and business requirements are volatile. But, pervasive computing can produce a profound effect on the applications or the products that they make. Also, it will have a positive effect on organizing and managing the business.

1.2 Technical Building Block

In the previous section, we presented a vision of the pervasive computing system. To build a pervasive computing system, there are important elements needed, which are hardware, software, business model, etc. Looking at the development of the pervasive system, the technical block consists of following crucial elements:

1. Mechatronics system (sensors and actuators)
2. Radio-frequency identification (RFID)
3. Internet of things (IoT)

Each building block is discussed in the following sections.

1.2.1 Sensors and Actuators

The sensors and actuators are important elements of the control system needed in the pervasive computing environment. Figure 1.2 represents a typical mechatronics system that encompasses the sensing unit, actuating unit and controlling unit that will work in the real world.

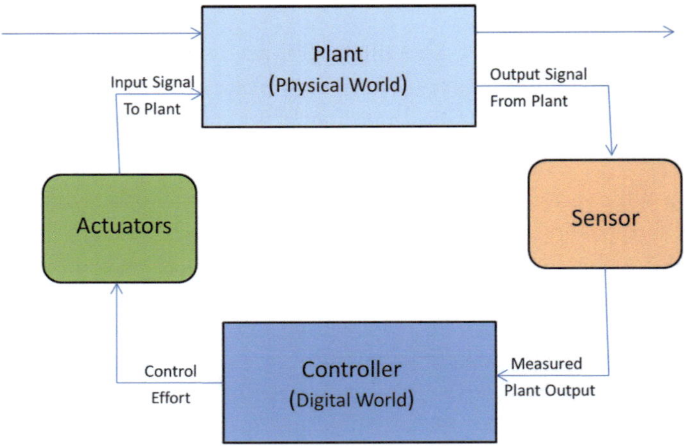

Fig. 1.2 A typical view of mechatronics system

This system having sensors and actuators is called mechatronics system. The sensing unit consists of sensors and other required components like filters, modulators and signal conditioners. Actuators are the part of the actuating unit along with optionally coupling mechanism and power supply. The controlling unit mostly lies between the sensing unit and an actuating unit. The control unit reads the inputs (information) from the sensing unit, performs the required operation to make decisions and produces the outputs to the actuating unit.

Sensors

This is a device that produces a proportional output signal (magnetic, mechanical, electrical, etc.) when that device is exposed to the physical incident or aspects (force, displacement, temperature, etc.). The sensors sometimes also called transducers. The main purpose of the sensor is to read a change and respond to the changes in the physical phenomenon or incidents. The job of the sensor is to read one form of energy and then convert it into another form. For example, the thermal sensor reads the change in the temperature and converts it into electromagnetic force (i.e. proportionate change in the electric energy). There are several types of sensors available and are used for specific purposes. The classification of the sensors is represented in Table 1.4. This table consists of mostly all sensors including the latest sensors like nano-, micro- and smart sensors.

Selection of Sensor(s)

Looking at the types of sensors discussed in the above section, it is challenging to decide which sensor shall be selected or used for a specific application. There is not a single formula or parameter that helps to decide the sensor. We are presenting Table 1.5 that shows the list of factors that individually or collectively decide the sensor type or sensor suits to the application.

Table 1.4 Types of sensors

Sr. No.	Sensor type	Explanations
1	Temperature sensors	These sensors are used to measure the temperature with a wide range (−100 to 200 °C). They are versatile sensors and cheaper too. The example includes thermistors, thermal transistors and thermocouples
2	Acceleration sensors	These sensors are used to measure the frequencies with a high range (40–45% of natural frequency). These sensors are highly sensitive, rugged and compact. The example includes a seismic and piezoelectric accelerometer
3	Pressure, force or torque sensors	These sensors are used to measure static as well as dynamic force. This could be used to measure the pressure also. The precision of the measurement is really high by these sensors. The example includes a strain gauge, tactile sensor and piezoelectric load cells
4	Proximity sensors	These sensors are used to detect any object present nearby you with no physical contact. It looks for the changes in the field. The example includes capacitance, photoelectric, inductance and Hall effect
5	Flow sensors	These sensors are used to determine the speed with high rates. This can be used for upstream as well as downstream measurement. The example includes rotameter, ultrasonic type, pitot tube and electromagnetic flow meter
6	Linear or rotational sensors	These sensors are used to detect the movement of the objects. The example includes electrical tachometer, optical encoder and Hall effect sensor
7	Light sensors	These sensors are used to measure the intensity of the lights. The high sensitivity measurement is the aim of these sensors. These sensors capture the digital image of the surroundings/fields. The example includes photodiodes, photoresistors and photoconductors
8	Smart material sensors	These sensors are used in various applications like robotics to measure strain, to measure humidity, to measure the change in the viscosity, etc. The example includes optical fibre (as a strain sensor, a force sensor, temperature sensor, etc.), magnetostrictive sensors, etc.
9	Micro- and nanosensors	These sensors are used to measure small things like flaws in the pipes or proximity between the blood cells and catheter. The example includes micro-CCD image sensor, micro-ultrasonic sensor, micro-tactile sensor

Table 1.5 Selection of sensors

Sr. No.	Factor	Remark
1	Range	This is the parameter where the requirement for an application looks for the specific range of the sensed parameter. It could be the minimum and maximum acceptable values
2	Accuracy	This parameter is vital in most of the applications. This value is the difference between the true or expected value and the measured value
3	Precision	There are certain cases where the application requires to see the accuracy and its occurrence over time or iterations. If the accuracy needs to be to observed frequently, then the precision parameter is vital
4	Sensitivity	This parameter is useful where the change in output is observed whenever there is a change in input. This parameter gives you the ratio of change in output when the input is changed
5	Response time	Looking at the agility of the application, the response is checked. This parameter is the time interval between the input given and output produced
	Bandwidth	This is useful in several applications where the output magnitude drops like by 3 dB
	Signal-to-noise ratio	There are applications where the errors are caused due to the noise. This parameter is the ratio between the signal magnitude and the noise measured at the output

Actuators

Actuators are the power behind mechatronics systems. It is a component which is responsible for controlling and moving the system or mechanism. They require energy and control system. They produce the change in the physical environment or system (motion, force, heat, etc.) after receiving the control commands/energy (electrical signals). With the help of power supply along with a coupling mechanism, the actuators are used in the system. Basically, two motions provided by the actuators are linear and rotary. The types of actuators are represented in Table 1.6. In general, two types of actuators are binary and continuous. This classification is based on the number of output states. A binary actuator is like relay which has two stable outputs, whereas the continuous actuator is like a stepper motor. Depending on the type of energy, the actuators are classified. The classification will be hydraulic, electric, electromagnetic, electromechanical, etc.

Selection of Actuator(s)

Due to the actuator's effect on the dynamic behaviour of the system, the selection of actuators is more complex than the sensor selection. The coupling mechanism and power supply are crucial in the selection of actuators. The coupling mechanism may not be needed if the actuator gives an output that can be interfaced with a physical system directly. In general, the parameters that need be considered in the selection of actuators are represented in Table 1.7.

Table 1.6 Types of actuators

Sr. No.	Actuator type	Explanations
1	Electric actuators	These actuators are used electronic type. They are consuming low power and had a high-frequency response. The example includes diodes, a bipolar transistor, thyristor, etc.
2	Electromagnetic actuators	These actuators have a large force and work for a short duration. They have an on/off control mechanism. The example includes solenoid-type devices, relay, electromagnets
3	Electrical actuators	These actuators are used to control the speed by voltage or current. These actuators have high efficiency, fast response and high speed. The example includes a DC motor, AC motor and stepper motor
4	Hydraulic and pneumatic actuators	These actuators are appropriate for line movement, produces high horsepower and no electric shock danger. The example includes Air motor, valves, cylinder hydraulic motor, etc.
5	Smart material actuators	These actuators are suitable where small motion with high frequency and high motion with low frequency are required. This includes electrostrictive and piezoelectric, magnetostrictive and electrorheological fluids
6	Nano- and microactuators	These actuators are used in the micromechanical system. It can use any smart material. The example includes microvalves, micromotors, and micropumps

1.2.2 RFID

After the study of sensors and actuators, the next important technical block is radio-frequency identification (RFID) technology. The detailed description of RFID is given in the following section.

RFID is one of the successful technologies that have been found in the several applications at small as well large scale. This identification technology is automatic and effective for a diverse set of objects. These objects could be artefacts (manufactured or natural), humans, locations, media contents and the services offered by mobile computing. Due to the flexibility and fast responsive solution, the RFID has become tagging and identification technology.

Table 1.7 Selection of actuators

Sr. No.	Factor	Remark
1	Range of motion	The range of motion is either linear or rotary. Based on this parameter, the actuator is selected
2	Continuous power output	This parameter is checked when the maximum force achievable continuously the temperature in the limits
3	Accuracy	This is another parameter that decides the selection of the actuator. The relation between input as well as output is monitored, and it should be linear
4	Heat dissipation	How much heat in wattage is dissipated in the continuous operation decides the actuator?
5	Speed characteristics	The relationship between the speed and forces decides the actuator
6	Frequency response	This is another parameter that checks the range of frequency on which the output is followed with the input. This is mostly required in case of linear actuators
7	Power requirement	This could be important parameter sometimes when the actuators are selected based on the type of power (DC or AC). The number of phases along with the level of voltage and capacity of current

Overview

'Identification Friend or Foe' (IFF) was the first application of RFID, and it was in World War II by the British [27]. The purpose was to decide whether to attack or not. The use of RFID in the military is continued. 'Electronic Article Surveillance' (EAS) was the first commercial application to prevent theft in the system [28]. This was based on the concept of tags that used to store a single bit. The set and unset setting were done based on the customer's entry and exit from the store. If this bit was not unset while the customer is an exit from the store, then this system generates the alarm to prevent theft. The prominent application was found by the USA and Norway, when RFID was used in the collection of tolls on roads [28] .

In the supply-chain and tacking domain, RFID has been used widely and effectively [29]. This has continued its effectiveness in task automation and personal tracking [30].

RFID technology consists of several parts. The RFID tags (transponders) consist of the following parts:

- Microchip
- Antenna
- Case
- Battery (in case of active tags only)

The size of the tag depends on the frequency and the area of use. This could be a millimetre or the size of a book. It uses Automated Data Collection (ADC) technology. To transfer the data between the reader and other objects, it uses radiofrequency waves. This is very fast and does not require physical contact between the tagged item and reader (scanner). The cost of the components in the RFID is low cost. The component of the RFID system is depicted in Fig. 1.3.

Since the serial number does not give information to the user and also not useful to keep the track of the items, the combination of RFID and the backend proceeding will help to do some useful computations like the serial number associated with the product information in the backend.

The RFID system consists of the main components:

- RFID tag
- RF antenna
- Network
- Workstation

The tagged item (RFID tags) is scanned by the RFID reader. The reader or scanner has to power the tag continuously. This information is being sent back to the backend for further processing. The backend is a database and the application interface. Using Ethernet, the reader sends the information to the workstation for processing purpose.

Sources of Energy

In the view of energy or power, the three types of RFID tags are:

1. Passive

   This tag does not have a power source internally. Hence, they depend on the power or energy persuaded by the reader. This means that the reader must be set its field until the transaction is completed. These tags are smaller in size and cheaper in the cost. The range of these tags is 2 mm to some few metres. The lifespan of these tags is unlimited.

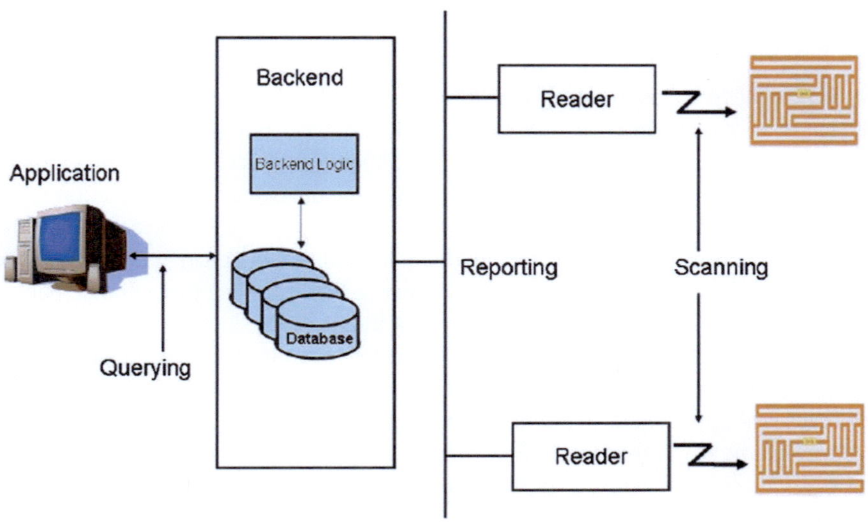

Fig. 1.3 Components of an RFID system

2. Semi-passive
   This is the second type of tags, and they have an internal energy source that keeps the chips powered on. As the chips are powered on, these tags are faster responding to the queries. The reading range is large than the passive tags, and hence, they are used widely than passive tags.
3. Active
   This is the third type of tags, and like semi-passive tags, they have an internal energy source. In this tag, the energy is supplied to generate a signal to the antenna and to power the chip. These tags send signals without being asked. The range of these tags is up to tens of metres. This helps to locate the objects. The life of these tags is up to 5 years.

Authentication
The aim of authentication is to confirm that a thing is what it claims to be. This is used to separate the authentic readers from other readers using tags. This is achieved with the help of encryption with key management (preshared). The lightweight encryption protocols can be used like AES for the authentication process. In the context of a pervasive computing environment, the RFID can be used as a technical solution which is cost-effective.

1.2.3 Internet of Things

This is the third technical block of pervasive computing. During the research, the development of RFID has directed to the new technology. In 1999, the RFID development community has coined the term Internet of things (IoT). Due to the increase in mobile devices, embedded communication, cloud computing and data analytics, IoT has become more practical technology [30].

Definition of IoT: It is a collection of physical objects. The Internet is not only about the network of computers but also the network of devices of any size, any type, vehicles, smartphones, toys, cameras, home appliances, medical instruments, animals, buildings, people, etc. All are connected, communicated, sharing information using the suitable protocol to achieve tasks like tracking, positioning, smart reorganization, real-time monitoring, personal care, upgrade, administration, etc. [31]. However, the definition of IoT results in an interaction between the following three things: person to person, person to things/machine, things to things or machine to machine.
Vision of IoT: The main vision of IoT is to create an environment in the diversity of devices which are connected to each other using wired or wireless connection that achieves new applications or services for a common goal. This is creating a world by converging digital, real and virtual world that will make transportation, cities, homes, energy and other areas more intelligent [32]. The goal of the IoT is to empower the things to be connected anytime, anywhere, at any circumstances with anything and anyone using network and services, and it is described in Table 1.8.

Table 1.8 Characteristics of IoT

Sr. No.	Characteristics	Explanation
1	Connectivity	This characteristic of IoT enables the devices or things to access the network easily. IoT provides common ability to interact, access and share information easily from any device to any other devices
2	Dynamic nature	The states of the things in the network are easily handled. The states could be connecting or disconnecting from the network, sleep or active mode, etc. IoT also handles the context very well and that can be time, speed, location, motion, etc.
3	Interconnectivity	The IoT enables the infrastructure to connect any other infrastructure and global information system
4	Heterogeneity	The diversity of the devices (size, type, power, hardware, OS, etc.) is well considered, and they can interact with other network or devices using the platform(s) provided by IoT
5	Safety	Though everything is connected to everything, the safety of the things and the services provide by the IoT are secured with the help of security measures. This includes the safety of devices, personal information, endpoints, networks, moving objects, etc.
6	Scalability	The devices are growing at a faster rate; however, IoT is capable to handle scalability in terms of devices, services, the volume of information, etc. The increasing numbers of devices and their seamless communication are well managed in IoT
7	Things-oriented services	IoT is capable of providing services based on the things and its context. The services are provided within the given constraint such as semantic consistency and privacy protection among the communicating entities

1.3 Issues and Challenges

Having a discussion on the pervasive computing overview and technical building block of pervasive computing, this section presents the issues and challenges in pervasive computing. The issues are presented in three ways:

1. Design issues
2. Technical issues
3. Security issues.

1.3.1 Design Issues

This section focuses on the design aspect of pervasive computing. This includes the following challenges.

Pervasive Interaction
The user interfaces are functional or task oriented. The system takes the user to what he is allowed to do. However, in pervasive computing, the preferences are given on user task rather than technology. Hence, the issue is interaction mode and that includes task-oriented interaction, multimodal interaction and human interaction pattern.

Scalability
As the number of devices and its variance is growing at a large scale, the issue is how to handle that scale. The challenge in pervasive computing is the device scalability. The communication in the pervasive system is anyone-to-anyone. This implies that the user or devices can communicate with any other user or devices at a time. Hence, the user and service scalability are the challenges need to be considered in the pervasive system.

Adaptability
The challenge is how to make adaptive information display. The interaction is not a single mode. It would be challenging to consider flexible voice-based commands (input) and respective output.

Context Awareness and Adaptation
Context is defined in the form of location, user activity, environment, capabilities, etc. The challenge in pervasive computing is tracking the location of a user or device. This could be an indoor or outdoor location. The challenge, in this case, is the accuracy and the limitation of available infrastructures.

The mobility is another concern in the pervasive system. The challenge is how to detect moving objects continuously. The objects can be a user who is walking, running for a bus, travelling in a metro, etc.

The user environment is another challenge where the system should behave accordingly. It could be a meeting room, theatre or sitting alone at a remote place, etc.

1.3.2 Technological Challenges

This section specifies the technological challenges considering the pervasive computing system. In broadway, the challenges are discussed in two parts: hardware and software.

Hardware Challenges

High-resolution flat-panel display is the requirement in the pervasive computing system. The devices and the services need hardware that is effective in nature, and interaction with the user should follow the interface golden rules.

The speed of the hardware should be fast enough that pertain the characteristics of pervasive computing. The speed should match the response rate for the application or the services that it offers.

Though the storage is not an issue looking at the cloud computing; however, the challenge still persists in the pervasive system as it is connected with a large number of communicating entities and collecting/sharing a large amount of information. The challenge is the bandwidth of wireless networks in pervasive computing. At a time a large number of users with their increased numbers of intelligent devices are interfacing with the network; hence, the bandwidth should be large enough. Last but not least, challenge in this category is the power consumption. The hardware should consume low power during computation in the pervasive environment. The challenge is to develop high power capacity hardware consuming less power.

Software Challenges

Though the basic Internet architecture and protocol stacks are working with the current scenario, it does not suffice the requirements of the pervasive computing system. The first challenge is the dynamic configuration of the hardware or software in the pervasive computing system. The mobility is the main concern as the devices are not stationary all the time.

Hence, the challenge is how to address device mobility. On the same line, the users are also not stationary in pervasive computing. The users are moving, walking, running, etc. The state of the user is dynamic. Hence, the user mobility is another challenge for the designers of pervasive computing.

The challenge is the application execution in the heterogeneous computing environment as well as the heterogeneous environment. The migration of application in the heterogeneous environment is another challenge that has to be addressed very well.

The 'technology-hidden' approach of pervasive computing is the important feature of pervasive computing. The transparent linking in wired and wireless networks is the main challenge that should include short and long wireless communications.

Implicit interaction with the user in the pervasive computing environment is another challenge where augmenting user's ability in the context of texts.

1.3.3 Security Challenges

This section focuses on the security challenges in pervasive computing.

As we mentioned in Sect. 1.1.3, pervasive computing has significance over the business and its economics. The pervasive computing makes the organization to enhance its horizons and go global that crosses organizational boundaries. The interaction will cross single and multiple organizations that will expose the possibilities of attacks.

In pervasive computing, each and every element is necessary. Hence, there are the chances of hacking everything which is worth to hack. Hence, there is a challenge to design and develop 'out of box' security measures.

The difference between the traditional security mechanism and the pervasive computing security mechanism is that it should be adaptive and context-aware. This is the main challenge to provide security to the system.

The communication between the pervasive computing elements is small. The chances of stealing and losing the data are more in those communications, e.g. biometric authentication.

The trust is the important aspect in the pervasive computing that has been valued in a specific way. As the experience and the recommendations are vital in deciding the trust scoring mechanism, the challenge is to maintain 'experience collection' and calculate trust.

1.4 An Overview of This Book

The smart objects and the intelligent computing systems are assisting users to make their jobs easier. The vision of pervasive system is to create an intelligent environment that includes the smart objects and the applications for the user. This book presents detailed description on pervasive system. At the outset, it is important to understand the progression of pervasive system. This book presents the fundamental concepts and architecture of pervasive system.

The pervasive system consists of several building blocks and technologies. However, the context and context information are vital to understand the user requirements, response to the user request sent by understanding context, etc. This book gives the detailed discussion on context, context information, and its use in developing pervasive system. This book provides the features of pervasive system and also presents the challenges of the pervasive system.

Context-aware pervasive systems are capable to handle the user context and their impact on the desired applications. This chapter covers the important context-aware computing models. The context and its information have brought a value to the user services that help the business to grow. Hence, this book is useful for not only the users but also the businesses to increase their businesses in an exponential order.

This book brings the user modelling as important aspect to enhance user experiences, access to the information, etc. The profiling and personalization are the driving entities in the design of pervasive system. This book envisions the use of context-aware clustering which could form as a fundamental step in future Internet.

The book shades some light on the diverse environment, distributed architecture that readers can understand the standard reference model, and also on the issues and challenges in the pervasive system. The authors have presented context-aware algorithms based on machine learning. Considering the mobility of users or devices in the pervasive system, this book covers the important aspects about telecom domain. The telecom industry and the media context are no longer disconnected; hence, this book covers the telecom domain.

The book will also stand an appetite for the readers to understand the security challenges in the context-aware pervasive system. The key factors for secure access are privacy, security and trust. This book provides the need for secured context-aware pervasive, method for the risk assessment and high-level secure context management framework. The uniqueness of the book is that it provides the attacks, attack analysis and its modelling to understand the attacks in context-aware pervasive system. Another feature of this book is the proposed framework for privacy, security and trust. The security algorithm ensures the services offered to the users are protected in the context-aware pervasive system.

1.5 Conclusions

In today's time, the penetration of smart devices and their extensive use has changed the users' life. The technology has not only brought services to the users, but also a value to the businesses. This chapter has elaborated on the vision and objective of pervasive computing by representing the technology change from networking to a pervasive computing system.

The chapter has brought clear technical elements of the pervasive computing system that includes RFID, sensors and IoT. The discussion of types of devices (sensors, RFID) helps to decide which devices required in the given environment. Also, the selection parameters mentioned in this chapter bring us the scope of pervasive computing. The chapter has presented several issues and challenges of the pervasive computing system. Furthermore, the overview of the book is presented in the last part.

References

1. Weiser, M.: The Computer for the 21st Century. Scientific American, Sept 1991
2. Apple Inc.: Available at:https://www.apple.com/watch/
3. Garlan, D., Siewiorek, D.P., Smailagic, A., Steenkiste, P.: Project Aura: Toward Distraction-Free Pervasive Computing Pervasive Computing. IEEE (2002). Available at: https://www.cs.cmu.edu/~aura/docdir/aura-pervasive02.pdf
4. Massachusetts Institute of Technology, Cambridge, "MIT Project Oxygen 2001". Available at: http://www.ai.mit.edu/projects/oxygen/oxygen-book2001/oxygenbook2001.pdf
5. UCI Berkeley: The Social Side of Science: A Human and Community Endeavour. Available at: https://undsci.berkeley.edu/lessons/pdfs/social_side_of_science.pdf
6. Birrell A.D., Nelson, B.J.: Implementing remote procedure calls. ACM Trans. Comput. Syst. **2**(1) (1984)

7. Satyanarayanan, M.: A survey of distributed file systems. In: Traub, J.F., Grosz, B., Lampson, B., Nilsson, N.J. (eds.) Annual Review of Computer Science. Annual Reviews, Inc. (1989)
8. Gray, J., Reuter, A.: Transaction Processing: Concepts and Techniques. Morgan Kaufman (1993)
9. Davidson, S.B., Garcia-Molina, H., Skeen, D.: Consistency in partitioned networks. ACM Comput. Surv. **17**(3) (1985)
10. Borg, A., Blau, W., Graetsch, W.: Fault tolerance under unix. ACM Trans. Comput. Syst. **7**(1) (1989)
11. Needham, R.M., Schroeder, M.D.: Using encryption for authentication in large networks of computers. Commun. ACM **21**(12) (1978)
12. Cavoukian, A., Reed, D.: Big Privacy: Bridging Big Data and the Personal Data Ecosystem Through Privacy by Design. Retrieved 28 June 2014, from https://www.ipc.on.ca/images/Resources/pbd-big_privacy.pdf (2013)
13. Satyanarayanan, M.: Fundamental challenges in mobile computing. In: Proceedings of the Fifteenth ACM Symposium on Principles of Distributed Computing, Philadelphia, PA, May 1996
14. Royer, E.M., Toh, C.K.: A review of current routing protocols for ad hoc mobile wireless networks. IEEE Pers. Commun. **6**(2) (1999)
15. Bhagwat, P., Perkins, C., Tripathi, S.: Network layer mobility: an architecture and survey. IEEE Pers. Commun. **3**(3) (1996)
16. Brewer, E.A., Katz, R.H., Chawathe, Y., Gribble, S.D., Hodes, T., Nguyen, G., Stemm, M., Henderson, T., Amir, E., Balakrishnan, H., Fox, A., Padmanabhan, V.N., Seshan, S.: A network architecture for heterogeneous mobile computing. IEEE Pers. Commun. **5**(5) (1998)
17. Noble, B.D., Satyanarayanan, M., Narayanan, D., Tilton, J.E., Flinn, J., Walker, K.R.: Agile application-aware adaptation for mobility. In: Proceedings of the 16th ACM Symposium on Operating Systems Principles, Saint-Malo, France, Oct 1997
18. Flinn, J., Satyanarayanan, M.: Energy-aware adaptation for mobile applications. In: Proceedings of the 17th ACM Symposium on Operating Systems and Principles, Kiawah Island, SC, Dec 1999
19. Weiser, M., Welch, B., Demers, A., Shenker, S.: Scheduling for reduced CPU energy. In: Proceedings of the First USENIX Symposium on Operating System Design and Implementation, Monterey, CA, Nov 1994
20. Lebeck, A.R., Fan, X., Zheng, H., Ellis, C.S.: Power aware page allocation. In: Proceedings of the Ninth International Conference on Architectural Support for Programming Languages and Operating Systems, Nov 2000
21. Zhang, J., Zhu, M., Papadias, D., Tao, Y., Le, D.L.: Location-based spatial queries. In: ACM SIGMOD'2003, San Diego, California, USA, 9–12 June 2003
22. Ward, A., Jones, A., Hopper, A.: A new location technique for the active office. IEEE Pers. Commun. **4**(5) (1997)
23. Spreitzer, M., Theimer, M.: Providing location information in a ubiquitous computing environment. In: Proceedings of the 14th ACM Symposium on Operating System Principles, Dec 1993
24. Voelker, G.M., Bershad, B.N.: Mobisaic: an information system for a mobile wireless computing environment. In: Proceedings of the Workshop on Mobile Computing Systems and Applications, Santa Cruz, CA, Dec 1994
25. Katz, R.H., Long, D., Satyanarayanan, M., Tripathi, S.: Workspaces in the information age. Report of the NSF Workshop on Workspaces in the Information Age, Leesburg, VA, Oct 1996. Available at: http://www.cs.berkeley.edu/~randy/NSFWS
26. Weiser, M., Brown, J.S.: The coming age of calm technology. In: Denning, P.J., Metcalfe, R.M. (eds.) Beyond Calculation: The Next Fifty Years of Computing. Copernicus (1998)
27. Songini, M.L.: Wal-Mart details its RFID journey. Computer World, Mar 2006
28. Wikipedia, the free encyclopedia: Available at: https://en.wikipedia.org/wiki/Electronic_article_surveillancehttps://en.wikipedia.org/wiki/Electronic_article_surveillance, Sept 2019
29. Smith, J.R.: RFID-based techniques for human-activity detection. Commun. ACM **48**(9) (2005)

30. Infocomm Media Development Authority, Singapore "The Internet of Things (IOT)": Available at: https://www.ida.gov.sg/~/media/Files/Infocomm%20Landscape/Technology/TechnologyRoadmap/InternetOfThings.pdf
31. Dr. Vermesan, O., SINTEF, Norway, Dr. Peter, F., EU, Belgium: Internet of Things—From Research and Innovation to Market Deployment. River Publishers' Series in Communications (2014)
32. Dr. Vermesan, O., SINTEF, Norway, Dr. Peter, F., EU, Belgium: Internet of Things—Converging Technologies for Smart Environments and Integrated Ecosystems. River Publishers' Series in Communications (2013)

Chapter 2
Context and Context Management

Abstract The user and user requirements are the main important concerns to design a pervasive computing system. The context of a user or device is another essential element in designing a context-aware pervasive computing system. This chapter is focused on what is the context, the variety of contexts, and the impact of that context information in developing a context-aware computing system. This chapter has also discussed several context-aware computing modelling as a part of the state of the art. On the line of existing models, this chapter has proposed 8C—a new context-aware computing model. The chapter also describes how RFID and sensors are used to sense context and shares the context information to further layers of context-aware computing system architecture. This chapter concludes with the context, its interpretations, storage and other important aspects.

2.1 Context Overview

The trends towards pervasive computing are accelerating as the computing devices and services are going beyond the laptop and desktop environments. The users are moving from desktop to mobile devices having diverse interfaces and environments. The user is not limited to a specific environment or device. The users may have choices to select several devices to access many services. These devices could be fixed or mobile. The devices are getting manufactured with enough capability to provide diverse services with a flawless interface. Another important aspect is that the users now have to access diverse services in a diverse environment from diverse devices. The challenge is to provide an interaction model that can handle device diversity. Another challenge is recognizing and understanding the computing environment where the application needs to be used effectively. Hence, there is a need to consider the adaptability between the two important elements, i.e. users (devices and applications) and environment (electronic and physical). Figure 2.1 represents important elements and their interactions.

User experience is enhanced as the computing devices and applications automatically adapt to changes in the user's environment (physical or electronic environment). Figure 2.1 represents the interaction between users and environment. Users are using

P. N. Mahalle and P. S. Dhotre, *Context-Aware Pervasive Systems and Applications*,
Intelligent Systems Reference Library 169,
https://doi.org/10.1007/978-981-32-9952-8_2

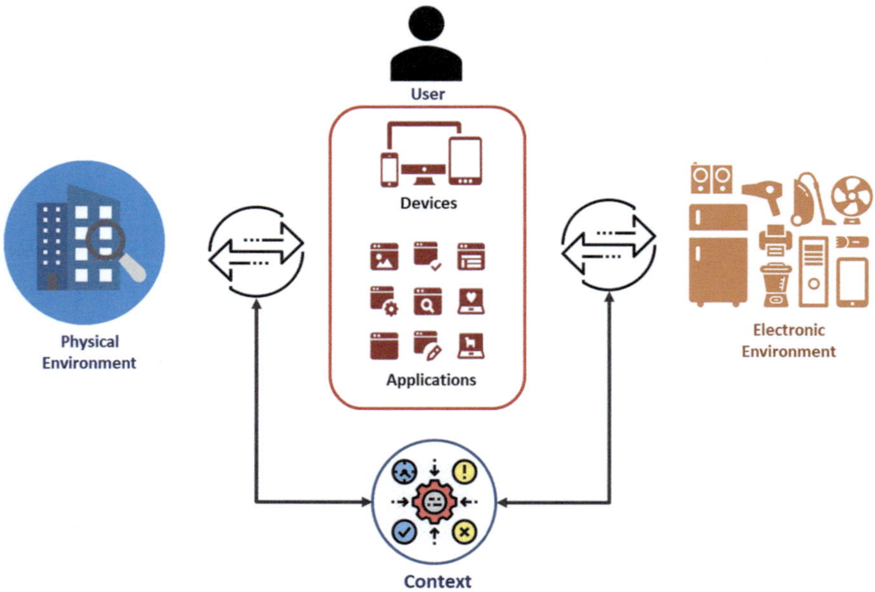

Fig. 2.1 Context: interaction between the user (devices and applications) and environment (physical and electronic)

the applications or services on their diverse devices. The interaction between the devices and applications is vital, and it depends on the users surrounding the environment.

The interaction between the user's physical and electronic environment creates valuable information which helps to define the context. The information generated during the interaction between the users, applications and the environment is called as 'context'. For instance, a user is browsing a news channel or watching a movie on TV at home reveals useful information. Another example, a user is searching for a shopping mall near to him while driving his car after leaving his office. The useful information could be the user's location, time, interest, etc.

However, the challenge in pervasive computing is to consider the context and its usage during selecting the services offered to the users. The use of context and impact of context-aware applications requires an understanding of context, its classifications and its computing.

2.1.1 What Is Context?

In general, a context can be any information that describes a situation associated with the communication between users, services and the nearby environment. However, there are several ways the context has been defined. The information generated during the interaction between the users and the environment. Having said before, the users in the environment (physical or electronic) access the services. This interaction generates information that collectively forms a context.

According to Webster's Dictionary, a context is 'an entire situation, environment or background related to some character or happening'. This definition seems quite general in regard to context-based computing. This section describes the definition with examples. In 1994, Schilit et al. [1] have defined context as identities, the location of objects and people. Also, context describes the changes to objects. The location of users and their devices reveal information to understand the customized services to be offered. These services should be automatically adopting the changes in the user's location.

To extend the above definition, in 1997, Brown et al. [2] have defined contexts like location, daytime, temperature, season and user identity. The recommendation system has evolved based on the user's choices and preferences. Capturing the user's location or temperature gives additional information to the service provides to understand the user's need and their preferences. In the same year, Ryan et al. [3] have included users' environment, location, time and identity as context.

In the year 1998, Dey et al. [4] have contributed to the definition by considering user's location, orientation, time and date. Also, the author has also considered user's attention and emotional state as a context. Even though these definitions are defined, the challenge is how to apply. It is unclear from the context definition that how information is context or not.

Looking at the above definitions, none of the definitions has talked about user preferences or interests. Also, most of the definitions said the context as a situation or environment. But the question is which environment? Is it users' or applications' environment? Since the definitions from Brown et al. [2] and Franklin et al. [5] said context is users' environment, whereas Harter et al. [6] and Rodden et al. [7] mentioned about application's environment. On the other side, Hull et al. [8] have emphasized users' current situation is the context.

The two definitions by Schilit et al. [9] and Dey et al. [4] are closer in terms of operational aspects. These definitions are focusing on not only the users but also the user's location, environment or surrounding, nearby resources. As per their definition, the context is execution environment which is prone to change. The environment could be one or a combination of the followings [4, 9]:

- Physical environment: This includes the noise level and lighting conditions.
- User environment: This includes several parameters like users' location, set of nearby people and social state.
- Computing environment: This includes input and output devices available for users, network connectivity and its capacity, processors, computing cost.

This shows that the definition of the context should include users' various states like physical, emotional, social and informational states. Looking at the above definitions, the detailed definition of context is defined by Dey et al. [4]:

> any information that can be used to characterize the situation of entities (i.e. whether a person, place or object) that are considered relevant to the interaction between a user and

> an application, including the user and the application themselves. Context is typically the location, identity and state of people, groups and computational and physical objects.

This definition is more precise, and it is in line with the previous definitions discussed. With the help of capturing context, it is easy to know what the user is trying to do in the given state. Hence, finding users' objective is challenging but the context can help to identify it.

Context helps us to do things in a better way. To observe or perceive information, the context gives influences. Based on the context, we can decide how to observe information. There are many sources used to sense the context. However, context gives the inspiration to identify the method to observe or gather information. A context also helps in managing a large amount of information surrounding an object. This limits the additional requirements to handle a large amount of information. Looking at the amount of information generated, it is a challenge to understand useful and not useful information. Context helps to determine the importance of the information. Another important feature of context-aware computing is adaptability. The context helps to adopt the application according to the environment. For example, the music application at home changes according to the user's mood.

The model of identifying context required input and allowing the application based on it contributes to the context-aware model. The model should be able to consider both inputs like implicit and explicit inputs.

However, the definition has focused on acquiring information, but it does not reveal the way of acquiring the information, i.e. manual or automatic or both. But, looking at the ideal case of a ubiquitous computing environment, the information acquiring method should be automatic. But, in the real environment, user intervention is required since all the context cannot be gathered automatically. This definition has emphases on the information that helps to identify the categories of context.

2.1.2 Context Classification

As we know that speaking the same talk at one location may give the same result at another location. For example, sharing a joke at one location may give a laugh but the same joke at the funeral does not produce a laugh. It is important to understand the information shared or exchanged during the communication among the users or between machines or users and machines. Context is playing an important role in the design and development phase of application or system.

Looking at scenario 1 and scenario 2 mentioned in Sect. 1.1.1, the '*Swayam*' needs to identify various types of information about the environment to handle the queries or difficulties faced by *Mr Kiran* and *Ms Anjali*. This information could help to perform the operations correctly. This information could be the user's location, the interest of the user, the problem faced by the user, connecting users' devices to the available network, fetching users' schedule, the location of other communicating entities, etc.

After observing the scenario, there is a diversity found in the context information. This diversity helps to assess the context and categorize it. The important entities are identified as users, things and locations. The users could be an individual or group of users. The users may be at the same location or distributed in an environment. The things could be either physical objects or software component. For example, the thing could be a smart device or application or file. Lastly, the location could be the geographical places like the airport, presentation room, buildings, offices and streets.

The context and its information are of different types. Generally, the context is classified as primary context and secondary context [10]. The information collected from sensors is called primary context. This information is without fusion and does not involve the use of context. On the other side, information that uses data fusion is called secondary context. In short, the secondary context is generated from the primary context. For example, primary context is the name of a person and his email id, and a phone number is a secondary context.

Basically, there are four categories of the context as follows:

- **Physical context**: This type of context deals with physical entities/values. The examples in this category include temperature, traffic conditions, noise levels, lighting.
- **User context**: This type of context deals with the users and information related to the users. The examples are users, their profile, the users' locations, etc.
- **Computing context**: This type of context information is network connectivity, sources like display, workstations, printers. The computing context can be communication bandwidth.
- **Temporal context**: This type of context information is related to time. This includes season, year, month, week, day and time of the day.

Along with the above categories, time can be also context information. This includes the year, month, week and day. These categories are important to perform useful computation. The computation can identify the nearby objects (proximate selection), a process to add, remove or altering the existing components (automatic reconfiguration of context), producing different results according to the context change (contextual commands and information) and rule-based adaptation of context-aware systems (context-trigger systems).

Based on the definitions, scenarios and important entities, the categorization of context information is digital identity, location, time and activity [4] (Fig. 2.2).

The digital identity is an electronic representation of a real-world entity. Digital identity represents 'who' is the entity in the environment. The digital identity could be the online equivalent of an individual. For instance, online users are characterized or described in terms of various things like personal data, personal assets, online behaviours, user interests, social relations. The identity could be a combination of above-mentioned things. An attribute is a distinct and measurable property belonging to a particular entity. An attribute could be a piece of information. The union of attributes forms an identity in the digital world. The complete set of characteristics that defines you is called identity. The set can have a first name, last name, mobile number, device ID, etc.

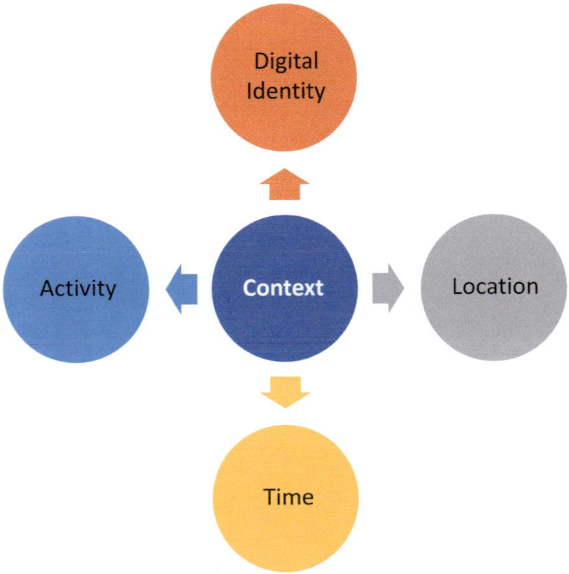

Fig. 2.2 Context classification [4]

On the other side, the identifier is an attribute or set of attributes of an entity which uniquely identifies the entity in the given context. The identifiers in the offline mode could be user passport, identity cards, driving licences, etc. The identifiers in the online mode could be login or sign in details, certificate or digital certificate. In the namespace, the identifier should be unique and should be used by the application(s).

A location is position information about the object or people in the 2D form. This includes elevation and orientation. The location can be all the information that helps to describe the relationship between objects like containment, proximity or co-location. It specifies the geographical area in the given active state. This area is circularly formed by latitude and longitude (centre point) and the specific radius from the centre point. The active and inactive states of the location are decided by the location provider. The location provider could be a GPS or network. GPS provides the most accurate position of the object. Based on the specific context, GPS is used. A network can use Wi-Fi or cell towers to identify the location of the device. The requirement of this network is the availability of a data network and consumes less power. As compared to GPS, a network provides less accurate location details. The location is also applicable to places too.

To characterize a situation or moment, time is an important context. It allows us to leverage off the value of historical information and its richness too. However, the important note about the time is that the time is used in association with the other context like time span or timestamp. This association will help to retrieve and gain contextual information to decide the level of relevance. For instance, the time reveals additional information when the user is in an active state and inactive state. Also, the patterns are identified based on the user's behaviour over time.

Lastly, the actor finds essential characteristics of the entity that can be recognized. The activity or the state could be the temperature in a room, ambient light level in a restaurant or noise level on a street. The activity could be referring to a person and its physiological factors tiredness or vital signs. An activity associated with a person could be reading (a book), watching (a movie), talking (to friends), etc.; for a person, the activity could be associated with a group of persons. Description of group activity can be attending a lecture or having a meeting. The software components activity could be uptime of CPU, state of files, state of the application, etc.

The information produced by each category mentioned just now can help to get an additional piece of information (inferred context information) and can assess the situation. From the user's identity, a piece of information could form another context information. For example, the phone number or area code could reveal some information from users' address proof document. The complex information could be retrieved from the set of context information. For example, by knowing the number of students in a classroom, their relative positions in a classroom and in a classroom occupancy, and if they are doing something (talking or listening, not talking, etc.), one can determine if a lecture is taking in that classroom.

The context information and additional information from the collective contexts (inferred context information) can help to determine the context-aware services or applications and its effective use in the computation. To shed some light on context and its use, we now move to context-aware computing.

2.1.3 Context-Aware Computing

Having presented a discussion on context, definition and its categories, it is now important to observe the use of context information in the application or computations. To understand the context-aware computing, let us consider another scenario where the communication is purely based on the context.

> *Mr Chaitanya is a reviewer for a conference where Ms Yashita has submitted a research paper. Mr Chaitanya has completed a review of a paper for Ms Yashita and he informs his 'intelligent assistant' to discuss the review comments with her. As Ms Yashita is busy in attending the practical session, her 'intelligent assistant' informs Mr Chaitanya to contact Ms Yashita by email or meet in person after two hours. On the other side, 'intelligent assistant' of Mr Chaitanya see the schedule of Mr Chaitanya and found that he is busy during the time slot suggested by Ms Yashita's assistance. Therefore, Mr Chaitanya's assistant drafts an email to Ms Yashita and share it immediately.*

The above scenario consists of two persons (a reviewer and an author) and an intelligent assistant. The communication among these entities depends on the information about entities, event, communication mode, etc. For each entity, the information about the other entity is a must. Also, the relationship between these entities is vital for making decisions. The communication mode is also important based on the state of the entity.

Such information is very important to perform computation on behalf of another entity present in the given environment. Also, the source of information is important to collect information. The source of information may be users, device owners and communication modes/channels. Some sources having the sensors also. Some information is gathered from different sources (user schedule and a user's context) and collectively used for the computations. Hence, there is a need to discuss the properties or characteristics of context information which are shown in Fig. 2.3.

- Static and dynamic: The first property of context information is static or dynamic. The pervasive system could have non-variant information of entities in the environment like family relations or date of birth. However, the pervasive system is designed for the dynamic nature of context information. This information is changing on a large scale from time to time. The information are user location, preferences or choices, actions, etc. The collection of static information is easy compared to dynamic information. Mostly, the users are the source of static information, whereas the sensors act as the source of dynamic information.
- Imperfect: The information in the pervasive system is incorrect if it does not match with the real instance. The information is not correct if its context is not identified well. Also, the information source does not provide information on time, and then it could lead to imperfection in the pervasive system. Due to the dynamic nature of pervasive computing, the information is being collected by sensors and stored in a repository. The change in the context may lead to inconsistency in the context information. The hardware or software failure can lead to the defective values of context information.
- Many forms (polymorphism): There are many examples where the context information given by sensors is represented in one format that an application does not support or asked for another format. The temperature sensors give the temperature in degree Celsius. However, the application wants it to represent as high, medium

Fig. 2.3 Classification of context information

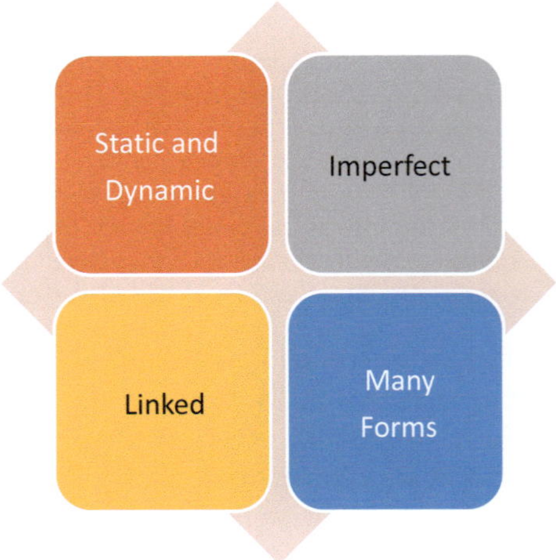

or low temperature. Hence, there is a need to develop a context model that must support numerous representations of context information.

- Linked: The information collected may have a relation with the other context information such as the devices, users and communication channels which are related in one or the other way. This implies that the context information can be generated or derived from the other context. Based on the user's location, information collected over time can be used to identify the users' action.

Having presented some of the key characteristics of context information, now we present the use of context information in the computation. The classification is based on context and use. The context can be used as a visualization or presentation. The presentation means to show the information about the context as and when required. On the other side, context information is used to trigger some actions. So, the classification of context-aware computation is *presenting context information, service execution and context information attachment for retrieval.*

In the first type, the context information is presented to the users. With the help of an application context, information is collected and the same is presented to the users. This context information helps users to take actions. There are several examples from the literature that demonstrate this type of computation. One example is navigation or map where the user's location or user vehicle's location is represented. This simply gives information to the users. Based on the location information, the user may take further action, but context-aware computation does not give specific action. However, the map also shows nearby sites that user is interested [11–17]. In 1994, Schilit et al. [9] have shown the example of this computation type where printers, projectors, etc., are suggested to the users based on the user's location. Presenting information on the display to the user is also an example of this category.

The second type of computation is service execution. This type of context-aware computation represents not only the context information but also provides actions to be performed. As per the context, commands are triggered to the user in a system. Similarly, on the behalf of the user, the system is recognized and responded according to the context information. In 1992, Want et al. [18] explained the 'Teleport System' that shows how users' screen environment moves as the user moves from one machine to another machine. This allows the system to maintain the state of work despite the change in the machine or working space. Similarly, Want et al. [18] represented the car navigation system. The route is displayed as the user has set the source and destination. However, in case the user misses the defined route then the system shows the correct route on the screen. Hinckley et al. [19] have explained how mobile devices change the setting as the user requirement or context changes. Based on the context, the mobile devices receive reminders or notifications on the user's device. The location information can help the application to give reminders to the users. For instance, if the user is in an area, then the application can suggest the list of friends near that user can meet and say hello. Also, there could be another example where the system reminds a user to drop letters as soon as that user is near a postbox [20]. One more example is the phone sound or profile setting. As soon as the user enters into a meeting room, the application changes the sound setting and changes the profile from loud to silent.

The third category of context-aware computation is attaching context information. This includes the set of applications that tag apprehended data to the relevant context information. There are many examples where the information collected is useful to enrich the value of context information. For example, if a user has taken a tour in a zoo, then the information collected during the tour can be attached to the context like the location of the zoo and time spent at the zoo. [21]. Similarly, the meeting conducted between some people can reveal some additional information. The additional information could be the location of the meeting, time and number of attendees of the meeting, etc. Another research work 'Placeless Documents' proposed by Rhodes et al. [22] attach context to desktop files or network files. The attached information helps the owner of the file for retrieving the files in the easiest way.

2.2 State of the Art on Context-Aware Computing

With the rapid use of mobile devices, the popularity of the pervasive computing systems has been increased exponentially. These devices are used seamlessly in the user's life. Mobile and sensor devices cooperatively work together to satisfy the user's requirement. The traditional method gathers the data and processes in a computing network. However, this traditional method leads to some problems. Hence, including context and context-awareness could resolve those problems to a high level. Tiny devices like users and their requirements are the central focus of the application. There are many applications and categories. With the help of context and context awareness, the system is capable to adapt their services according to the context and avoid user intervention. Therefore, the effectiveness and usability increased by considering context and its awareness into account.

To develop and implement context-aware applications, there should be necessary components like

- **Sensors**: The element is used to sense and gather the context.
- **Widgets**: The sensors are interfaced with the help of software called widgets.
- **Interpreters**: The context, context information and its abstract information are provided by the interpreters.
- **Discoverer**: The discoverer is the application on the top of an interpreter who has access to any of the components.

There are some basic steps involved in developing context-aware systems or computing. These steps jointly called as context-aware computing models. In the existing models, the basic steps are

- Collection of context information,
- Aggregation of collected context information,
- Reasoning the collected context information.

Table 2.1 Context-aware computing models [23]

Sr. No.	Model name	Description
1	Location-aware model	As the mobile devices are increased, the services are offered based on the user's location on those devices. Hence, the context-aware systems are looking to locate the users, nearby devices or environment to provide location-based services. Hence, location information is useful to develop a context-aware system that uses magnetic card readers, cameras, mobile towers, GPS, satellites, etc.
2	Object-oriented model	This type of models deals with objects that represent various contexts. The physical and virtual entities are used, represented and included in the processing
3	Logic-based model	A context model is represented using facts, rules and expressions. This model is adaptive as it adds, removes or alter the facts based on the developed logic
4	Ontology-based model	This model is purely based on thoughts, ideas, etc. The description of ideas and their relationship is considered in this model

Based on the above steps, some models are presented here [23]. The context-aware computing models had been proposed by many researchers in the past. Few of them are represented in Table 2.1. These models are the models that consider the basic parameters and are specific.

The above models need to be extended as the set of applications are increasingly widely. Also, the communication and integration of the services are leading to a new model called an 8C model. This model is now discussed.

8C Model

This model consists of eight (08) steps for context-aware computing. These steps are Categories, Components, Connection, Communication, Cooperation, Convergence, Content and Context, and it is shown in Fig. 2.4.

1. Categories
   Having said before, the use of mobile devices and resource-constrained tiny devices (sensor nodes) has been increased a lot in the user's life. The combination of these devices has used extensively in several applications. Few applications are like home applications, agriculture application [24], healthcare applications [25] and military applications [26]. This category of applications deals with several components and the communication among themselves.
2. Components
   There are many components involved in forming real-time, context-aware and efficient services to the users. The components like RFID, sensors and sensor networks are resources in the pervasive computing system. These devices or components are embedded in the computing environment such that the interaction with each other to provide useful information about the context.

Fig. 2.4 8C model

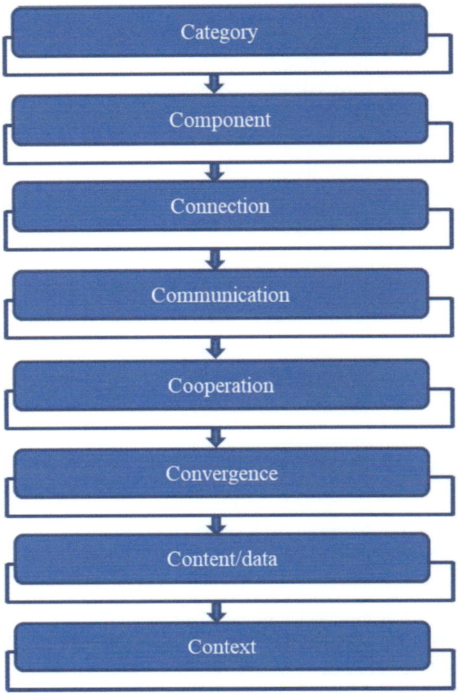

3. Connections
   The components are connected to each other by means of connecting technologies. The surrounding environment may be the same or different, but the connection is established between the devices and software entities.
4. Communication
   Based on the user's requirement, the components communicate with each other. The device like mobile or laptop communicates to the printer with the help of local connection. There could be some applications where smart cards need to be communicated to mobile devices to complete the task.
5. Cooperation
   The communication between the components may belong to the same network or different. However, they may communicate as a part of application demands. The cooperation among the components from the same or different networks is essential. The synchronous mechanism must be established to handle the heterogeneous components to work in cooperation. Also, the devices may be the same or different. The cooperation among them is equally essential for seamless communication.
6. Convergence
   The architecture of the Internet is inefficient to handle the set of diverse devices, homogenous and heterogeneous network, ad hoc networks, etc. The present Internet needs to be enhanced, and next-generation Internet should be capable of con-

verging different networks and services. The mixing of the different networks will lead to an effective and efficient pervasive computing system.

7. Content/Data

   The next-generation Internet will have several communications among different networks that lead to a huge amount of data. The contents are generated from several domains like agriculture, smart city and health care. Also, the customs services offered to the users lead to specific data generation. Hence, the content or data needs to be handled in a specific way.

8. Context

   This is the last and important step in the model which makes the pervasive computing system as a context-aware system. The context information is gathered, classified and used in the appropriate application to offer required services to the user.

Considering the steps mentioned in the 8C model, the context is sensed, collected, classified and used to provide services to the users. The context is sensed using several ways. The following section describes a few methods.

2.2.1 Using RFID to Sense Context

To reduce cost, increase efficiency and fasten the process of traditional industries, the RFID enabled applications are penetrating and widely used. It has been used in many industries like supply chain, inventory and logistics. [27, 28]. The application will be more fasten as the RFID has a low cost and easily combined with information and communications technologies (ICT). The RFID provides smart devices that are capable to identify, monitor and track the objects.

In the context-aware pervasive systems, the environmental events and conditions, users' intents and their current situations are sensed by technologies like RFID. In short, the RFID has the capability to sense internal as well as external events and conditions. These internal and external conditions form its context. The sensing subsystem like GPS, camera, sensors recognizes the external and internal events and conditions. On the same line, identification facility provided by RFID can sense external and internal contexts. For example, user location-based services are provided [29] and users' intents are read to provide required services [30].

The diversity of RFID-based context-aware pervasive systems is increasing, and it has diversity in nature. Hence, it is important to discuss what context is collected, used and provide context-aware applications. There are several RFID systems presented in the literature that used the tag values to gain the context and decide the corrective usage action. The different tag values are shown in Table 2.2.

The contents and the formats of the services are determined by the context. The information from tags is collected and used for analysis. A change in tag value determines the user's requirements or surrounding conditions. With the help of RFID features, the context is sensed using RFID.

Table 2.2 Tag values and meaning [31]

Sr. No	Tag value	Meaning	Usage
1	User	Owner of the tag	User ID, profile, preference, role, etc.
2	Object	Physical entities having tag attached	Object number (ID), monitoring and tracking
3	Location	Place where a tag is positioned	Follow-me, location-based queries (LBQ) or location-based spatial queries (LBSQ)
4	Neighbourhood	Association among connected tags	Coordination among tags, monitoring, etc.
5	Behaviour	User's intention	Observation, monitoring

The value read by tags is more accurate and explicit. In contrast, the values read by sensors are not clear because of noise or jitter. Also, the speed of concurrent reading data is faster compared to other sensors. As the tag values and their categorization are stored in the database, there is no need to have complex algorithms to retrieve tag values.

The tag and reader are the two important elements of RFID. During reading a tag value, another important value read is the reader's ID. This ID acts as a location identifier and user identifier. Another important feature of RFID is that it can be combined with the sensor nodes. To gain physical values from the environment having smooth coordination with the sensor nodes, there are many systems where RFID has been used effectively. Bravos is a system that provides services based on the user's role [32]. If a user wishes to know trips near to his locations, then a city tour guide is one of the solutions [33]. User's activity is recognized and recorded using iGlove [34]. Another system has been proposed that stores the user's context and experience using life experience blogging [35].

2.2.2 *Using Sensors to Sense Context*

Wireless sensor networks (WSN) provide continuous monitoring of objects, environments, etc., in various fields like military, residential and health care. The advantage of WSN is that it is easy to deploy in an existing IT infrastructure in a cost-effective way. Integration of wireless devices with medical sensor nodes is much easier. With the help of components of WSN (actuators), the critical tasks are carried out. Most of the context-aware pervasive computing uses WSN which has the following essential parts:

- To identify the sensing parameters in a filed,
- To identify the sensing location and implementation of WSN,
- To transfer the collected data from file to the central station for decision-making,

- To analyse collected data.

The context is sensed by the sensors that are embedded locally. Once the data is sensed by the sensors, the software collects the data for further processing. There is no other layer in between the sensors and software to process received data. The sensors are not only sensing hardware but, in some cases, they provide useful information. Based on the data collection, the sensors used for sensing context are classified as follows [36].

Physical sensors: The physical sensors are those sensors which are being widely used. At the present time, there are several hardware sensors available which are proficient to sense nearly any physical data. Some examples of physical sensors are light sensors, audio sensors, motion acceleration sensors, location sensors, touch sensors and temperature sensors [37].
Virtual sensors: This includes sensors that track the user's location using but the location of users is also identified using virtual sensors. The working of virtual sensors is based on the applications or the services. For example, a camera of a safe home security system tracks the location of the owner in the home. However, using the home owner's location can be tracked in the home using his calendar application. The virtual sensors help to track a user's activity by means of keyboard input and mouse movement.
Logical sensors: This sensor is a combination of physical and logical sensors that use some additional information (from databases) to resolve issues in a system. One example to represent the use of a logical sensor is detecting or finding user's location. To trace the location, user's login to laptop or desktop, and device mapping into the database is analysed. The context-aware system can be visualized as several layers. Each layer is shown in Fig. 2.5.

The first layer is a sensor layer that consists of various sensors. The sensors could be hardware devices or any other sources which provide valuable information about context. The sensors are used based on their use to get the context. For example, to get audio data, microphones are used. Similarly, context and name of sensors (represented in bracket) are video context (camera), location (GPS, GSM), acceleration (motion sensors), temperature (thermometers), light (UV, IR sensors), physical attributes (Blood sensors, biosensors), etc.

Fig. 2.5 Layers in context-aware system [38]

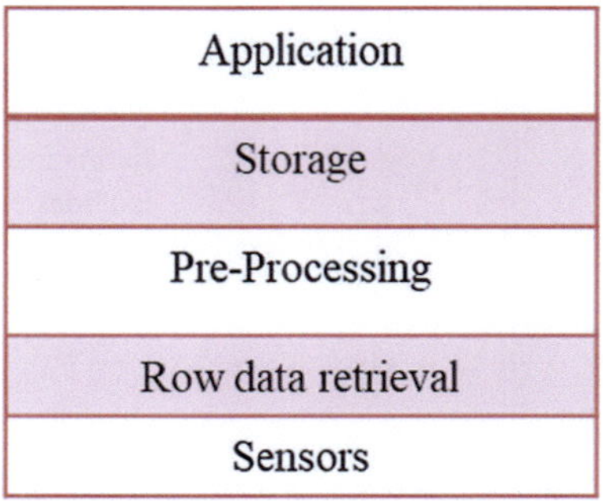

Once the context is sensed by the sensors, the second layer is row data retrieval. This layer uses drivers for physical sensors. For logical and virtual sensors, this layer uses APIs. The next layer is used for pre-processing which removes unnecessary data sensed by the sensor layer. The storage and management layer store the necessary context information and give access to the users. The last layer is the application layer which reads the events and instances are implemented.

2.2.3 Ontology-Based Modelling

As the context is sensed by the RFID or sensors or other sources, it must be stored for further processing. There is a need for a structure or model that stores the context information in a machine-readable form. The structure is called a context model. To cover possible context, it is challenging to develop a useable and flexible ontology. Strang et al. [39] have presented approaches for context modelling. The models are based on the way of representing context information and their exchanges. Few of the models are discussed in Table 2.3.

Along with the above-discussed models, there is one more model to represent the context. This model is ontology-based modelling which uses ontologies. In this model, the concepts and their association or relationship are described in this model. This modelling is very promising modelling to model the context information. The modelling allows to interface effectively. Also, the interfacing (reasoning) and high effectiveness are achieved in this modelling.

Table 2.3 Types of context models (modelling methods) [39]

Sr. No.	Model	Description
1	Key-value	This is the simplest model used to represent the services using a key-value combination of content
2	Object-oriented	In this modelling, the context types are represented as objects (location, temperature, etc.). The basic features of object-oriented (encapsulation, polymorphism, etc.) are used in the modelling
3	Machine scheme	Context and their attributes are represented in a hierarchical way (markup tags)
4	Graphical	To model context, a graphical representation is used in this modelling. The context data is modelling using uniform modelling language (UML)
5	Logical	In this modelling, the high degree context is represented. This includes expressions, facts and rules to decide to the model of context

2.3 Requirements for Representing and Exchanging Context

It is important to consider the perception of application developers during the design of a framework that supports all application tasks. The conceptual framework should support common applications in an automatic way. The designers should deal with the context and consider the requirements for the development of the framework.

The following requirements are essential to develop a contented framework development. The requirements are:

- Separation of context,
- Context Interpretation,
- Resource discovery and
- Context storage.

Each of these requirements is explained.

2.3.1 Separation of Context Use

Having said before on the definitions of context and its use, there is no specific way to handle context and its acquisition. Hence, the use of context is not found in all the applications. Therefore, the application developers are free to the easiest method for implementation. This section describes the two common ways to handle context. The first method is relating sensor drivers to the applications, and the second technique is the use of servers to deal with sensors.

As mentioned in [40, 41], the sensor drivers are used to deal with context and the drivers are completely linked or to the applications (hard-wired connection). In this method, the designers are required to write code that transacts with the drivers of sensors. The code is written using the protocol that suits or set by those sensor drivers. Even though this method is simple but has majorly two drawbacks. The first drawback is the difficulty in understanding and dealing with the context and the acquisition. Hence, the task of developing context-aware applications is difficult for application developers.

In this method, the good practices of software engineering are not followed. There must be a clear distinction between the low-level understanding of context and the applications. This makes a clear loss of generality. The reuse of sensors in other applications seems difficult, and their use in the parallel applications seems impossible.

In a real sense, the context should be handled in the same way the user inputs are handled and treated. The application developers should be able to handle inputs using user interface tools. The interface tools will allow the developers to use inputs without thinking on how the input was gathered. Hence, the abstraction is provided in terms of an interface that makes the interaction between the input and application. Hence, it is called as interactor or widget. The abstraction has been used in standard mouse and keyboard interaction. There are many examples where the widget abstraction is used widely like pen and speech input [42], virtual reality [14]. Such abstraction helps to separate low-level understanding of context, input and the semantics of the application. This means the code will be independent despite the use of any type of devices. The reusability can be achieved with the help of providing several instances of the same widget. With the help of a common external interface, an application can have the same handling method to all the widgets.

To obtain the context from sensing devices, there are several systems that support notification mechanisms [43]. However, the mechanism of common interfaces for the servers is the utmost importance. Hence, each application will deal with different servers in a different way. To get the flexibility, receiving context at one time may be useful for applications and on other time it may add some value to the applications. In short, once the context information is received by the applications, then it is important to understand to know whether the context is the same, or it has been changed. Further, how important is a change in context is useful to the applications. This will help the applications to decide and set the conditions for notification mechanisms.

The applications can handle the context information by differentiating the context, its type, its acquisition methods and its use. This also forces the application to work without worrying sensors and related context acquired from it.

2.3.2 Context Interpretation

The application can get the context information from the servers with the help of notification mechanisms and querying procedure. The context information is reached to the applications via several layers. The acquisition procedure could be simple or may be complex. For example, the notification should be given whether the presentation is started or not. Similarly, the location information can be used to interpret the location of users and identity information of users will let us know their co-location. Also, the scheduling information and sound information will identify the presentation is started or not.

The context information goes through several layers before it reaches the application. Hence, these layers should be known well to the designers of the applications and it should be transparent. Therefore, to achieve transparency, it is important to interpret the context before is used. The architecture of the application should consider the reusability of context interpretation. The recursive interpretation applications are proposed by several authors like [44, 45].

2.3.3 *Resource Discovery and Context Storage*

Even though the context, its use and its interpretation are important, the resource discovery and storing context are equally important. Before communicating an application with the sensor, it better to know the sensor (resource), type of information a sensor can give its location and its communication mechanism (protocol). In the distributed environment, it is necessary to know the sensor and it is running on which computer (hostname and port of a computer). Hence, the resource discovery is another requirement in the architecture as it should be hidden from an application. The resource discovery mechanism should be able to specify the context information required when an application starts. Such a mechanism will help to find ways to access the application components. For example, the application should notify whether the user leaves or enters a building instead of focusing on the use of sensing technology. The location information can be provided by discovering a location context resource.

To understand the trend and predict future values of context, the context storage can be used. To have historical information, constant availability is another requirement. The widget maintains little historical information. For example, file explorer keeps the most recent folders accessed and allows a user to select one of those folders in a quick fashion. It is up to the application whether to maintain little or complex history. Any component or devices that acquire the context should maintain the context history. If such context history is not maintained, then it is very difficult to do the analysis on context values. There could be the possibility of a device or component that collects the context which is not being used by any applications at some time. However, there could be another application, in future, may require that context information which is not being used by other applications. For example, the application could require knowing the location context in order to predict his future location. Hence, the application should support to store the context to provide historical information as well as future predictions.

2.4 Conclusions

The context and context information are the key elements to describe the situation related to communicating among the entities like a machine, user, surrounding environment. The environment could be a physical, user or computing environment. However, the context should include user's state that may be from physical to emotional to social, and informational.

This chapter has also presented various contexts and its type that includes identity, location and time. The necessary elements or components to develop context-aware applications are sensors, widgets, interpreters, etc. This chapter also discussed the context classification like physical, user and computing context. There will be designers to design the required context-aware applications.

This chapter provides a new model for a context-aware computing system that has focused on several elements from the real perspective on pervasive computing. This model discusses the possibilities of all the application and the process to sense the context from each application.

References

1. Schilit, B., Theimer, M.: Disseminating active map information to mobile hosts. IEEE Netw. **8**(5), 22–32 (1994)
2. Brown, P.J.: The stick-e document: a framework for creating context-aware applications. In: Proceedings of the Electronic Publishing '96, pp. 259–272. IFIP, Laxenburg, Austria (1996)
3. Ryan, N., Pascoe, J., Morse, D.: Enhanced reality fieldwork: the context-aware archaeological assistant. In: Gaffney, V., Leusen, M.v., Exxon, S. (eds.) Computer Applications in Archaeology (1997)
4. Dey, A.K.: Context-aware computing: the CyberDesk project. In: Proceedings of the AAAI 1998 Spring Symposium on Intelligent Environments, pp. 51–54. AAAI Press, Menlo Park, CA (1998)
5. Franklin, D., Flaschbart, J.: All gadget and no representation makes jack a dull environment. In: Proceedings of the AAAI 1998 Spring Symposium on Intelligent Environments, pp. 155–160. AAAI Press, Menlo Park, CA (1998)
6. Harter, A., Hopper, A., Steggles, P., Ward, A., Webster, P.: The anatomy of a context aware application. In: Proceedings of the 5th Annual ACM/IEEE International Conference on Mobile Computing and Networking (Mobicom'99), pp. 59–68. ACM Press, New York, NY (1999)
7. Rodden, T., Cheverst, K., Davies, K., Dix, A.: Exploiting context in HCI design for mobile systems. In: Proceedings of the Workshop on Human Computer Interaction with Mobile Devices, Glasgow, Scotland (1998)
8. Hull, R., Neaves, P., Bedford-Roberts, J.: Towards situated computing. In: Proceedings of the 1st International Symposium on Wearable Computers (ISWC'97), pp. 146–153. IEEE, Los Alamitos, CA (1997)
9. Schilit, B., Adams, N., Want, R.: Context-aware computing applications. In: Proceedings of the 1st International Workshop on Mobile Computing Systems and Applications, pp. 85–90. IEEE, Los Alamitos, CA (1994)
10. Perera, C., Zaslavsky, A., Christen, P., Georgakopoulos, D.: Context aware computing for the internet of things: a survey. IEEE Commun. Surv. Tutor. **16**(1), 414–454 (2014)
11. Abowd, G.D., Atkeson, C.G., Hong, J., Long, S., Kooper, R., Pinkerton, M.: Cyberguide: a mobile context-aware tour guide. ACM Wirel. Netw. **5**(3), 421–433 (1997)
12. Bederson, B.B.: Audio augmented reality: a prototype automated tour guide. In: Proceedings of the CHI'95 Conference on Human Factors in Computing Systems, pp. 210–211. ACM Press, New York, NY (1995)
13. Davies, N., Wade, S., Friday, A., Blair, G.: Limbo: a tuple space based platform for adaptive mobile applications. In: Proceedings of the International Conference on Open Distributed Processing/Distributed Platforms (ICODP/ICDP'97), pp. 291–302, Toronto, Canada (1997)
14. Feiner, S., MacIntyre, B., Hollerer, T., Webster, T.: A touring machine: prototyping 3D mobile augmented reality systems for exploring the urban environment. Pers. Technol. **1**(4), 208–217 (1997)
15. Fels, S., Sumi, Y., Etani, T., Simonet, N., Kobayshi, K., Mase, K.: Progress of C-MAP: a context-aware mobile assistant. In: Proceedings of the AAAI 1998 Spring Symposium on Intelligent Environments, pp. 60–67. AAAI Press, Menlo Park, CA (1998)
16. McCarthy, J.F., Meidel, E.S.: ActiveMap: a visualization tool for location awareness to support informal interactions. In: Proceedings of the 1st International Symposium on Handheld and Ubiquitous Computing (HUC'99). Springer, Heidelberg, Germany (1999)

17. McCarthy, J.F., Anagost, T.D.: EventManager: support for the peripheral awareness of events. In: Proceedings of the 2nd International Symposium on Handheld and Ubiquitous Computing (HUC2K). Springer, Heidelberg, Germany (2000)
18. Want, R., Hopper, A., Falcao, V., Gibbons, J.: The Active Badge location system. ACM Trans. Inf. Syst. **10**(1), 91–102 (1992)
19. Hinckley, K., Pierce, J., Sinclair, M., Horvitz, E.: Sensing techniques for mobile interaction. In: To appear in the Proceedings of the 13th Annual ACM Symposium on User Interface Software and Technology (UIST 2000). ACM Press, New York, NY (2000)
20. Beigl, M.: MemoClip: a location based remembrance appliance. Pers. Technol. **4**(4) (2000)
21. Pascoe, J., Ryan, N.S., Morse, D.R.: Human-computer-giraffe interaction—HCI in the field. In: Proceedings of the Workshop on Human Computer Interaction with Mobile Devices, Glasgow, Scotland (1998)
22. Rhodes, B.J.: The wearable remembrance agent. In: Proceedings of the 1st International Symposium on Wearable Computers (ISWC'97), pp. 123–128. IEEE Press, Los Alamitos, CA (1997)
23. Balavalad, K.B., Manvi, S.S., Sutagundar, A.V.: Context aware computing in wireless sensor networks. In: International Conference on Advances in Recent Technologies in Communication and Computing (2009)
24. Aqeel-ur-Rehman, Shaikh, Z.A.: Towards design of context-aware sensor grid framework for agriculture. In: Proceedings of World Academy of Science, Engineering and Technology, vol. 28, Italy, Apr 2008
25. Merico, D., Mileo, A., Bisiani, R.: Wireless sensor networks supporting context-aware reasoning in assisted living. In: Proceedings of 1st International Conference on Pervasive Technologies Related to Assistive Environments, Athens, Greece (2008)
26. Winkler, M., Tuchs, K.-D., Hughes, K., Barclay, G.: Theoretical and practical aspects of military wireless sensor networks. J. Telecommun. Technol. Available at http://www.nit.eu/publ/jtit
27. Weinstein, R.: RFID: a technical overview and its application to the enterprise. IT Pro, May/June, IEEE, pp. 27–33 (2005)
28. Harrop, P.: Rapid adoption of RFID in manufacturing and logistics sectors. Manufacturing and Logistics IT, Aug 2007. Available at http://www.logisticsit.com/absolutenm/templates/articlecritical.aspx?articleid=3190&zoneid=31X
29. Hsi, S., Fait, H.: RFID enhances visitors' museum experience at the Exploratorium. Commun. ACM **48**(9), 60–65 (2005)
30. Pering, T., Ballagas, R., Want, R.: Spontaneous marriages of mobile devices and interactive spaces. Commun. ACM **48**(9), 53–59 (2005)
31. Choi, J.: RFID context-aware systems. In: Turcu, C. (ed.) Sustainable Radio Frequency Identification Solutions, p. 356, Feb 2010. INTECH, Croatia, downloaded from SCIYO.COM. ISBN 978-953-7619-74-9
32. Bravo, J., Hervas, R., Chavira, G., Nava, S.: Modeling contexts by RFID-sensor fusion. In: Proceedings of Pervasive Computing and Communications Workshops, pp. 30–34 (2006)
33. Korhonen, J., et al.: mTag—architecture for discovering location specific mobile web services using RFID and its evaluation with two case studies. In: Proceedings of the Advanced International Conference on Telecommunications and International Conference on Internet and Web Applications and Services (AICTICIW), pp. 191–200 (2006)
34. Smith, J.R., et al.: RFID-based techniques for human-activity detection. Commun. ACM **48**(9), 39–44 (2005)
35. Cheng, Y.-M., Yu, W., Chou, T.-C.: Life is shareable: blogging life experience with RFID embedded mobile phones. In: MobileHCI, pp. 295–298 (2005)
36. Indulska, J., Sutton, P.: Location management in pervasive systems. In: CRPITS'03: Proceedings of the Australasian Information Security Workshop, pp. 143–151 (2003)
37. Schmidt, A., van Laerhoven, K.: How to build smart applications? IEEE Pers. Commun. **8**(4), 66–71 (2001)
38. Baldauf, M., Dustdar, S., Rosenberg, F.: A survey on context-aware systems. Int. J. Ad Hoc Ubiquitous Comput. **2**(4), 263–277 (2007)

39. Strang, T., Linnhoff-Popien, C.: A context modeling survey. First International Workshop on Advanced Context Modelling, Reasoning and Management, UbiComp (2004)
40. Harrison, B.L., Fishkin, K.P., Gujar, A., Mochon, C., Want, R.: Squeeze me, hold me, tilt me! An exploration of manipulative user interfaces. In: Proceedings of the CHI'98 Conference on Human Factors in Computer Systems, pp. 17–24. ACM Press, New York, NY (1998)
41. Rekimoto, J.: Tilting operations for small screen interfaces. In: Proceedings of the 9th Annual ACM Symposium on User Interface Software and Technology (UIST'96), pp. 167–168. ACM Press, New York, NY (1996)
42. Arons, B.: The design of audio servers and toolkits for supporting speech in the user interface. J. Am. Voice I/O Soc. 27–41 (1991)
43. Bauer, M., Heiber, T., Kortuem, G., Segall, Z.: A collaborative wearable system with remote sensing. In: Proceedings of the 2nd International Symposium on Wearable Computers (ISWC'98), pp. 10–17. IEEE, Los Alamitos, CA (1998)
44. Dey, A.K., Abowd, G.D., Wood, A.: CyberDesk: a framework for providing self-integrating context-aware services. Knowl. Based Syst. **11**(1), 3–13 (1998)
45. Kiciman, E., Fox, A.: Using dynamic mediation to integrate COTS entities in a ubiquitous computing environment. In: Proceedings of the 2nd International Symposium on Handheld and Ubiquitous Computing (HUC2K). Springer, Heidelberg, Germany (2000)

Chapter 3
Context-Aware Pervasive Systems

Abstract In the design of a pervasive system, the important element is the context. The context information has brought value to the user services and will exponentially increase the business in the coming years. The context and context management will help to understand the user requirements in precise and stands the heart-core element of the context-aware computing system. Having a precise discussion on the context and context management presented in Chap. 2, this chapter has focused on the context-aware pervasive system. The elements of the context-aware computing system are presented at the beginning of this chapter. Also, keen attention has been given on the context-aware network in detail in this chapter. Though the context-aware computing system is promising, it is coming with a few important challenges. This chapter ends with a detailed description of the issues and challenges.

3.1 Elements of Context-Aware Pervasive Systems

Overview of context, context sensing and the requirement to represent context have been presented in Chap. 2. The context is a situation that influences something to happen. In executing the service or application, the context is very vital and stands as a key role. Any system adapts their changes or behaviour to the present context with having less intervention of the users. Hence, the context is becoming the main role in the execution of applications. In Chap. 2, we also discussed the characteristics or properties of context. These properties are essential and become vital in the development phase of the context-aware applications. Some of the properties are shown in Fig. 3.1.

The capability of a system that understands the context and accordingly gives an anticipated response to the users is called 'Context Awareness'. Even though there are many systems considered as context-aware systems, the alternative names are an intelligent system, or smart system or adaptive system, etc. The context-aware system uses context and accordingly provides information or services to the users. The context is based on the users' task.

Depending upon the user interaction, the context awareness is classified into three levels as discussed in Table 3.1 [1]. The table consists of three types, namely per-

P. N. Mahalle and P. S. Dhotre, *Context-Aware Pervasive Systems and Applications*,
Intelligent Systems Reference Library 169,
https://doi.org/10.1007/978-981-32-9952-8_3

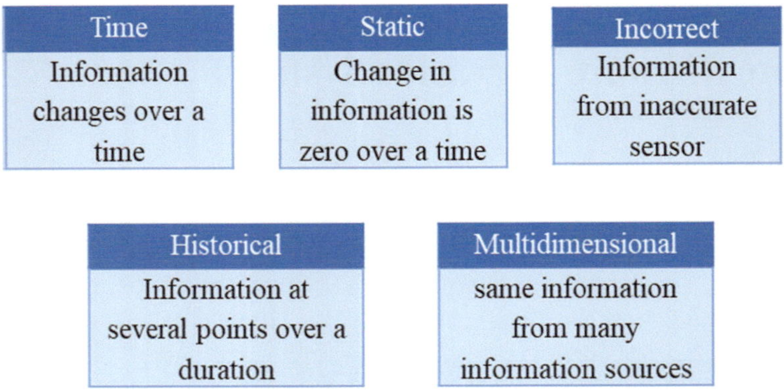

Fig. 3.1 A short summary of context properties

Table 3.1 Classification of context awareness [1]

Sr. No.	Type of context awareness	Details
1	Personalization	• A user sets the preferences to the applications or system in a manual way • For example, a user sets the type of music he likes when he comes home from the office. The music system set the given preference
2	Passive	• In this type, the environment is monitored continuously by the system. Based on the observations, user preferences are considered while taking decisions • For example, promo codes or discount coupons are shared to the users if the user is a frequent buyer of commercial products. This triggers the user to perform some action
3	Active	• In this type, the system monitors the environment and automatically perform the action without user intervention • For example, if an intruder has entered into a smart home, then the system automatically informs the landlord or the police department

sonalization, passive and active context-aware system. The classification is based on the involvement of the users in setting the preferences or choices in the application.

Profounder understanding of context is more important in deciding the type of context-aware system. The context-aware application needs to understand the relation between the entities involved in the system. Based on the complex relationship among the entities, Alegre et al. [2] have classified the context-aware system in two categories as execution context-aware system and configuration context-aware system.

In the execution mode of the context-aware system, the behaviour or the action of the system is based on the specific event or situation. For example, if a person

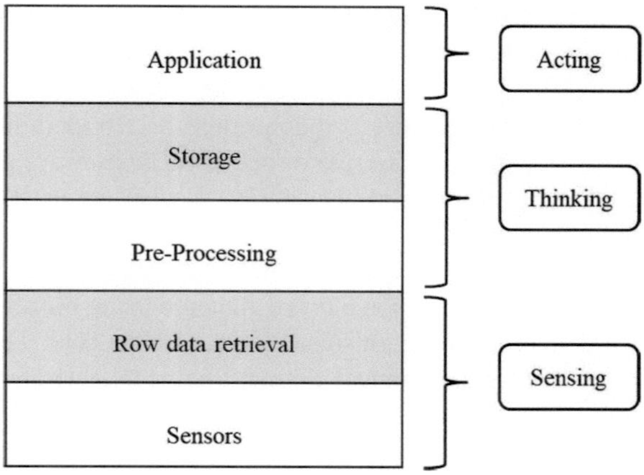

Fig. 3.2 Elements of the context-aware pervasive system

is attending a seminar and his mobile receives a call from his family member, then phone profile changes to silent mode automatically. In the configuration type, the system will make few adjustments in the behaviour or actions in the future. For example, the phone profile will be adjusted according to the user's preferences that will not turn the mobile phone into silent mode or answer the call.

We have seen that the context-aware pervasive systems can respond to any events, stimuli or to proactive inputs like artefacts or living things [3]. Some of these stimuli are software agents, robot, etc. Few parameters need to be considered in order to design and implement the context-aware pervasive system. The parameters include a number of users and devices, type of devices, user requirements, context collection methods. The important concern in the development of architecture context-aware pervasive system is flexibility.

In this regard, the primary task in the context-aware pervasive system is the identification and understanding of context and context-aware network. The identification and understanding of context are achieved by the architecture in the context-aware system. The layered architecture consists of several layers (see Sect. 2.2) where the sensor layer senses the information and passes on the information to other layers for further computations. Therefore, this section describes the essential components needed for context-aware pervasive systems that include sensing, thinking and acting. The mapping of layers and elements of the context-aware pervasive system is shown in Fig. 3.2.

3.1.1 Sensing

This is the first element of the context-aware pervasive system. Sensing is normally stated as the acquisition of information or data. This data or information is about the physical world which is being used by the pervasive systems. For example, the information could be temperature, location, etc. The information is unprocessed, and it is being sensed by the source of information like sensors. The main purpose of such sensing is to identify and perform useful actions. The action could be setting the sound system, brewing the coffee, changing the phone profile, etc. The sensing of information is vital as it contributes to the identification of context. The context helps in defining a passive or active context-aware pervasive system. To sense the context or information, the sensing devices are used. The sensing devices could be biological or non-biological or both.

The biosensor is another name to biological sensors that act as an analytical device. The biological components are gathered using biosensors. The component includes antibodies, microorganisms, tissues, nucleic acid. The integration of physiological transducer and biological elements helps to generate an electronic signal for detectors. The examples of biosensors are electrochemical, amperometric, blood glucose, thermometric, optical, etc. There are many applications where the biosensors are used for several purposes. Food analysis is one of the examples, where the information is sensed using biosensors. The sensed information is being used to decide the quality of food. Another example is medical diagnosis where several informations about tissues are collected for the numerous tests. The patients' health check-up (checking temperature, glucose in the blood) is effectively performed by the biosensors. The use of biosensors is also observed in other fields like biomolecules study, crime detection, field monitoring.

Non-biological sensors are used to detect electric signals and accordingly respond to it. The information gathered using non-biological sensors are comparatively huge and used in several application domains. The temperature is measured using contact and non-contact sensors. The information can be collected from several sensors can be motion sensors, IR sensors, touch sensors, light sensors, etc.

There are some issues with the sensors. Some of the sensors could be damaged which in turn produces a loss in data generation. Some sensors are implanted or surrounded by the real environment which requires high throughput or high accuracy. Hence, it is a challenge to decide the sensors and their sensing technology that work effectively for the context-aware pervasive systems.

3.1.2 Thinking

Once the information is gathered from the sensors using the sensing method, the next challenging task is to process it further to take desired actions. This task is called 'thinking'. This is a very critical component of the context-aware pervasive systems. With the help of gathering information from sensing devices, the next task is to analyse the information. The analysis of the information helps to give an appropriate response to the environment/user.

There are some examples where information is gathered, context is recognized, and appropriate action is performed. For example, if the person attending a meeting as per the schedule, the intelligent system should recognize it and corresponding action (turning the cell phone to silent mode automatically) is performed on the behalf of the person. The response to the system or user is based on the analysis of the information or user requirements.

There are two ways the information is being used. One way is to gain knowledge of the collected information. The information is pre-processed, and useful insights are captured to increase knowledge. The communication among the devices will generate information. For instance, temperature sensors in the premise will generate temperature information over a period. This generated information will reveal the temperature conditions or range in the given area.

The other way of using collected information is performing an analytical task to gain additional knowledge and predict future values. The patterns in the collected information present to get valuable insights. There are many examples where the technology used to understand the pattern of the information and reveal the faults or helps to take corrective actions. For example, in the electricity distribution system, the smart grids are used. Using computer networking and intelligence, the applications are developed to see improvement in the operations and maintenance of electricity. In a smart grid, every component can 'listen' and 'talk' and it is automated. In most of the cases, the customer calls the electricity department to inform electricity service interruption or out of service. However, in a smart grid, the department can see where the fault as each component of the system is sharing sensed data. The component is identified as a faulty component has stopped, sensing and sharing data. Also, the pattern of failure is identified by the analysis of information received from each component in a smart grid.

In some cases, both above methods are used to gain additional knowledge. The information from sensors is acquired in the context-aware systems and then compares it with other knowledge to gain further knowledge and is used to predict values. The important techniques used in the thinking phase are many, but to name a few of them are fuzzy logic, knowledge bases, neural networks, pattern recognition, etc.

3.1.3 Acting

From the above steps (sensing and thinking), the context information is gathered, and it is being analysed using many methods. Also, during earlier two phases the situations are recognized. Once the context information is sensed and collected, it is a turn to take desired actions on it to respond to the system or user requirements.

The actions are not only performance actions but also it may control actions. The control actions could be either stop action, cancel the action, or override action, etc. Looking at the same example of a smart grid, the control action could be stopping the electricity supply in a region due to faulty components. To give the responses to the control actions, the appropriate designs should be included in the context-aware pervasive applications. The performance actions deal with enhancing the actions to be performed on behalf of users. For example, incomplete or inaccurate information gathered by the sensing device will affect the performance and throughput of the system.

There are some applications that may need to investigate several types of actuators while interaction with users and the environment. This acting component is vital as it must concern with the appropriate actuators that must satisfy the requirements of users or the environment. These requirements are specific to the domain applications.

3.2 Context-Aware Networks

The recent development in smart devices, sensing technologies and context data management is providing information accuracy in both application and user environment. This information helps to identify the context and the complete view of the system. Also, the advancement in the sensor devices shows they are easy to embed in smaller devices in a faster way.

With the help of sensors and relevant technologies, the context-aware systems have emerged. These systems respond to the changes in the physical world effectively as per the users and environments requirements. The design of the context-aware system should not be only static. In another way, the context-aware system should be flexible and self-adaptive to sense, think and act as per the requirements of the operational environment. Hence, the combination of context-aware and adaptability design modelling of the system will be the best way to understand and deal with user requirements. The system should be dynamic, less human intervention, and reactive. Hence, the system should be self-adaptive to support effective abstraction and should be self-evolving at the time of execution. However, the issue is that the context changes dynamically. The movement of context in the context-aware system is shown in Fig. 3.3. This represents the life cycle of context discussed by Perera et al. [4] which has four parts.

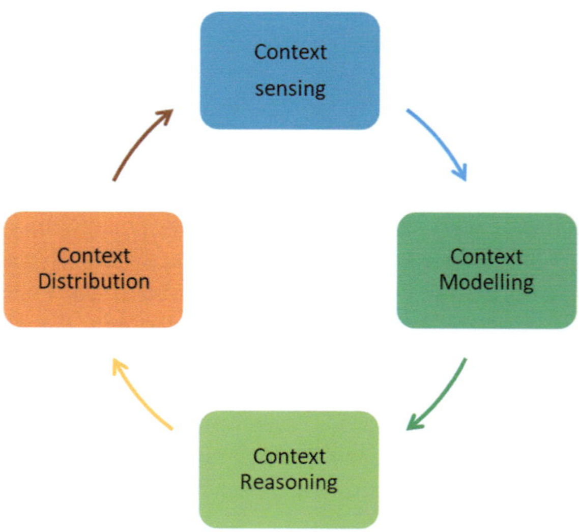

Fig. 3.3 Context life cycle [4]

The life cycle of context represents data movement from one phase to another phase. This cycle denotes the stage of sensing to the stage of its use. Each part is discussed now.

- Context sensing
  This is the first phase where context information is collected from the sensing devices. The sensing devices could be physical or virtual sensors. The information collection depends on various parameters as a source of information, its type, frequency, sensing method, etc. Once the information is collected, it is further passed to the next phase.
- Context modelling
  This is the second phase where the sensed data is modelled according to requirement. Here, the sensed data is represented in a way to understand the context. The representation of context reveals the characteristics/properties, meaning and relationships. Based on the requirements, the modelling can be performed using many techniques. Few techniques are key-value modelling, graphical modelling, markup modelling, logical modelling, ontology modelling, etc.
- Context reasoning
  This is the third phase where the pre-processing of sensed data and knowledge extraction are performed. For better understanding, the context is extracted. Imperfect and uncertain context is important concerns in this phase. Initially, the raw data is cleaned, missing values are filled, and data is validated. Once pre-processing is done, then data-fusion is performed to get more reliable accurate and dependable data. This is achieved with the help of combining information from several sensors. At the last, the context is identified using reasoning approaches (many techniques) like fuzzy logic, decision tree, hidden Markov model.
- Context distribution
  This is the last phase of the cycle where the context is given to the user. The distribution of context may use query technique. In this technique, the results are produced as per the user queries. The other technique is the subscription. In this technique, the results are given to the user periodically or where never the event has occurred.

The context-aware cycle is important to envision the use of context in real scenarios of a network.

3.2.1 Motivation

Having discussed before, there are many applications and devices that are supporting run-time environment and accordingly handling their behaviour as a response to users' requirements. Such applications or devices are increasing rapidly [5]. In the view of context awareness, a network should be reliable and stable. Also, the network should have control over the application (its creation and operation too). Hence, the context-aware network is a need today to support context awareness. It is a combination of the characteristics of a dumb network (Internet) and smart/intelligent network (like the telephone network). Having a discussion on the context and important elements of the context-aware pervasive system, following section, discuss the importance of context-aware network and their types. From the user's perspective, it is difficult to provide instructions to the complex systems. There are many reasons like understanding of the environment, high technical skills and continuous involvement of the users.

This is a motivation to understand the new ways of user interaction with the smart system. This smart system will work automatically and will respond to the applications on behalf of the user. The user's tasks are performed automatically. Such systems or network have several unique properties. The network that works dynamically and adopts themselves have properties like self-adaption, context awareness, proactiveness.

The proactiveness of context-aware networks has the capability to do communication and share information transparently. For example, in arranging a meeting, the users must deal with complex things like meeting agenda, time, location, number of attendees. They must be involved in several other things like selection process, calendar, no overlapping, desired technical skills.

The Director of the Institute receives a call from the Higher Authorities to prepare a Budget of the next academic year. As a reaction to the call, the Director would like to schedule a meeting with Heads of the Departments from the Institute. The professors from each department should have the details on the budget that includes the recurring and non-recurring amount.

Now, using the smart application, the Director select meeting scheduling application which enables him to add the participants in the groups. The Director also decides the time and location of the meeting using the same application. The notification to participants is sent on their mobile phones or laptops, or desktops via the same application and in turn, the budgeting task is initiated.

Each participant now communication at the time and location of the meeting. Each participant can see who is out of time, who is taking lectures, who is involved in other activity, etc. The optimal solution is calculated based on the availability of all the participants. The notification is sent to the Director about final meeting timings.

From the above scenarios, we observe that a network is dealing with many entities and several operations at a time. The entities are devices and users who wish to execute a meeting on the given task budget. Firstly, several tasks need to be identified. Also, the context needs to be understood to give responses to the communicating entities as per their requirements. The scenario represents the use of context-aware network where the tasks are identified, performed automatically on the behalf of the user. Some task needs manual intervention, but the network that runs the application can handle the task with the ease. There are many applications like 'automatic searching of the free time slot of a budget meeting with the Director' which is the strong motivation to understand the context-aware network.

Another interesting and frequently observed scenario is road traffic. The issue is being raised frequently, and there is a need for a supportive solution. Consider the following scenario in detail.

Mr Pravin and Mr Subhash are professionals who are working in an organization. They are frequent travellers from their homes to an organization to back home and covers a road distance 75 km per day. In a given scenario, they take 2 h of travelling daily. However, due to unavoidable and uncertain traffic conditions, they generally spend an additional 02 h. To address this issue, an "intelligent device" is connected to their vehicle as other commuters also have attached to their vehicles. This device captures valuable information about vehicle speed, location, and size too. The device locally calculates the traffic conditions like density, bottlenecks on the roads, alternate routes, trip timing, etc. Based on the communication with the other devices of other near vehicles, the application at Mr Pravin and Mr Subhash inform them about traffic situations and provide an effective route to reach them early at their destination.

In the above scenario also, the devices and the communication entities are performing and asking for some tasks. The intelligent device attached to the vehicle has envisioned a device that works in a context-aware network. Each device of the vehicle communicates locally. The locally means that the devices of each vehicle communicate with each other and exchange the information. This will fasten the process of feedback on the road conditions instead of depending on a global or central system.

Both scenarios demand smart mobile computing. The network combines the data from multiple sensors embedded in a network. The sensed data helps to understand the context; the meaningful measures are taken into consideration. These parameters or measures will assist in developing a context-aware network and intelligent applications that users aim to use them more often.

3.2.2 State of the Art

Looking at the examples mentioned in the motivation section of this chapter, there is a need for a system or network that understands the context, context-aware applications, its proactiveness, etc. There is the difficulty of such a system in the present state of the art as the system must consider several aspects like users, their device diversity, user needs, environment and it's setting, and other requirements too. There is no common model exist that has been agreed by the researchers.

The model should make changes in the architecture level or behavioural level to include self-adaptation property. The architectural solution is provided by Brun et al. [6]. The authors have proposed feedback looks that have crucial four activities. These activities are collect, analyse, decide and act. This unidirectional loop represents the inputs and the outputs of the states. However, the transmission of data among these states is completely omitted. In the first state, the data is collected from the environment sensors and other sources (internal or external). Once the data is gathered, then the next step is analysed. In this phase, the data is analysed according to the constraints and policies of the system. Also, user preferences are considered while analysing the data. This process could be complex as it involves errors and may have broken sensors. Once the system reaches from analysis to decide, the future predictions are done, and the action is performed which is the last stage. The similar feedback loop was used by IBM with the name MAPE-K [7]. This model uses a knowledge base that shares the data among the phases.

There are several opportunities for developers to develop new smart applications for mobile devices. As these systems are continuously analysing the context information, the surrounding environment is sensed, and knowledge is acquired. The awareness means a state that sense that system could is found. The sensing of context is performed by the sensors which are classified as physical, logical and virtual [8].

The widely used sensors are physical which provide the location data using RFID tags, GPS, GSM, Cameras, etc. The virtual sensors gather the context information from services and applications like operating system and network. The logical sensors receive information from other types of sensors and provide context information of a higher level. The event can also be an important element in the software cycle that could be one of three types: foreseen events, unexpected and expected events [9].

The proactive computing system 'Tennenhouse' was proposed by Tennenhouse et al. [10]. This system was a new mode computing that was leading towards human-supervised. This system has the capability to take the decisions acting on behalf of their users. The system can handle likely events whereas the context-aware applications and self-adaptation can handle the foreseen as well as unforeseen events. This makes the deviation of such systems from normal systems.

The state-of-the-art technologies are having few issues and challenges which include the self-adaptiveness and proactiveness. The new system should support mobility and context management. Another challenge is the security of the informed being exchanged in the environment.

3.2.3 Proactive and Self-Adaptive Networks

The users are surrounded by many devices like laptops, cell phones, smart watches, smart cares, smart refrigerators. Not only the use of devices has increased but also the mobile and wireless technologies are also being used prominently. The issue is the integration of the smart devices and intelligent services offered in a converged network. One of the solutions to handle convergence of networks is the automatic computing or a system that manage themselves. To set up a dynamic environment, intensive efforts lead to the development of middleware, architecture, prototype applications. These initiatives show the development of an intelligent network that can perform a complicated task. The network is classified into two types as a proactive and self-adaptive network. Each type is discussed now.

Proactive Networks

In the proactiveness network, the system controls the changes instead of waiting for the change or event to happen and then given reaction to it. The first proactive network was developed by Zampunieris in 2006 [11]. The proactive engine (PE) was designed to do proactive computing. This was the first system designed to run proactive rules. The proactive rule is defined as the reaction to the situation or the event detected in any situation. The proactive rule performs an action when there is a lack of event. For example, if an employee is absent and his theory session is not being conducted, then an automatic notification should be sent to the in charge of that department.

Coronado et al. [12] have proposed a proactive system called Learning Management System (LMS). This system has provided a structure that proactively defines the interaction between facilitators and learners. However, this system has limited interaction and collaboration. To enhance collaboration, a new mechanism is proposed by Dobrican et al. [13]. In this collaborative learning mechanism, the communities of practice have represented that allows proactive computing. The communities are formed according to the student's demographics. The three communities are students belonging to the same country, students from the same city and the students have the same program. This community in LMS allows to enrol the students in programs and to create their schedule and calendar, allowing chat room, etc., in an automatic and proactive way. Such systems avoid explicit interaction from administration authorities of LMS.

However, the proactive system or engines are single system and challenge is to handle the increasing number of devices and applications. Hence, multiple proactive systems can collaborate and offer many more services. Hence, there is a need for such a collaborative system that can communicate and perform actions together.

Self-Adaptive Networks

In an intelligent or smart system, self-adaptation is a strong motivation for researchers. The self-adaptive system not only manages or controls itself but also has a capability to adapt to the changes automatically as per the environment or user requirements. The self-adaptive system handles the behaviour of external environments along with internal elements and their observations too.

Krupitzer et al. [14] have proposed a taxonomy of self-adaptation, and it is represented in Fig. 3.4

Figure 3.4 represents important five dimensions which address the issues to satisfy the fulfilment of self-adaptation.

- Reason (Why to adapt?),
- Time (When to adapt?),
- Technique (What is the nature of change?),
- Level (Where to implement change?),
- Adaptation control (How to perform adaptation?).

The dimensions and related issues are discussed now [14].

1. Reason
   This dimension focuses on the reasons for change or adaptation. There are three reasons for it. The first reason is the technical resource change. For example, if the hardware is failed, software has errors. In both cases, the system has to adopt the change as the new hardware or software needs to be considered in the account.

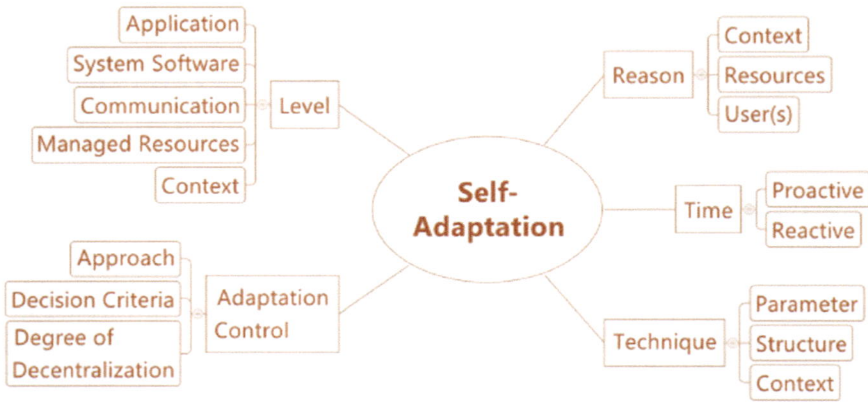

Fig. 3.4 Taxonomy of self-adaption [14]

The second reason is the change in the context. This is the environmental change that the system must consider. The third and last reason is the change from the user (trigger generated from the user). This could be because of a change in users' goal or requirements. The response of the changes needs constant monitoring of resources, context and user needs.

2. Time
   This is another dimension refers to the moment for system adaptation. The moment could be either before or after the need for a change. Generally, it is proactive or reactive. Among these two, proactive is preferred that avoids the flow of user in the system. However, the cost of handling reactive change is easier. Hence, reactive is considered in the self-adaptation system.
3. Technique
   Even though time is identified to adapt, but the important concern is what to adapt and what would be its nature? The types of techniques are classified as context, structure and parameter. The context change refers to the change in context. This could be changed from location to time to activity, etc. The structure change refers to the change in the hardware structure of the system. This could be the addition, removal or exchange of devices in the system. The last is the parameter that mentions the change in the system with the help of a change in the parameters. In some cases, the technique used to adapt to the system can involve a combination of all techniques.
4. Level
   This taxonomy refers to the change in elements like resources and logic. The managed resources are computational resources like laptops, smartphones, robotics. The giant resources like cars, production devices are also managed resources. On the other side, the logical adaptation includes the monitoring of resources, environments. This type of adaptation occurs at a specific level. For example, if the user is in a meeting then the smartphone application should change the profile to silent mode automatically. This type of adaptation is managed at the application

level. Similarly switching from one network to another network comes under communication level.
The logic of adaption is very crucial as it should know all the levels and the elements at each level. The awareness of levels helps to handle the adaption in a system.
5. Adaptation control
This is the last taxonomy. The adaptation control tells about how to do the adaptation. The design of adaptation control is based on three parameters, i.e. approach to set the change, criteria for decision-making, and degree of decentralization. The approaches could be internal or external. This approach is based on the separation of the additional layer (that has logical adaptation) from existing managed resources. The decision criteria are based on models, policies, rules, etc. The last approach depends on the size of the network and handling capability. If a system is small, then the centralized approach is effective. Otherwise, the system is decentralized based on logical adaptation.

In many programs or disciplines like software engineering or artificial intelligence, the study of self-adaptation has been a central attraction and helps to develop and implement intelligent software systems. Such systems are capable to comply with the run-time changes in an environment. In the state-of-the-art technologies, there are systems that focused on adaptation in its architecture [15] or implementation [16].

The important focus has been given to the change in the internal software system and changing implementation components of the software system without making any changes to the user interface. This means that the developers have to change the high-level abstraction without making any changes to the low-level implementations. One of the examples is a battery-aware architecture that has adopted the changes in the architecture to make it more efficient [17]. Another example is the adaptation of run-time multimedia media delivery hybrid application for smartphones [18].

3.3 Issues and Challenges

The context-aware pervasive system is though promising, but it has several challenges. Few of them are discussed in this section.

3.3.1 Routing

Like RFID, wireless sensor networks are important technology of a pervasive computing system. In the communication, the key aspect about the sensors is its range. Even the sensors are coming with high configuration, still the communication range and the network size are not matching. It indicates that the network size surpasses the range of the sensor nodes. Therefore, communication requires multiple hops through the message will be forwarded. Hence, the challenge is to route the messages in the pervasive system considering unreliable wireless communication and restricted properties of sensor nodes.

In the fast few years, the routing protocols have been proposed [19]. In some of the routing protocols, the route is selected based on the importance and priority of the message [20, 21].

However, context can be used to find the importance of the message. In the design of the routing protocol, the usage of data and its semantics can help to find the importance and priority of the message/data to be sent. There are several factors that define message importance. Some of the factors are the current context, the preference of the query, etc.

In the context-aware pervasive computing system, the context and context-dependent parameters will help to decide the routing of the message in the network. At every node, a knowledge base is stored that helps to perform computations are performed locally to know its importance. This computation helps the node to determine to forward or to drop the message. Using the semantic web and artificial neural networks, the knowledge base is created.

In the sensor network, the humidity or temperature acts as linear dimensions. These dimensions and their correlations help to estimate predicate selection in the queries. In sensor networks, the sensor nodes store dimensions based on the context. For example, the quality dimensions of temperate include all temperature values, time, region. Each characteristic is calculated, and its minimum and maximum values are stored.

The formulation of context values and their importance is done. A current context C is represented by its type c and associated value from a domain d. For example, the temperature is represented as:

$$\{C = C_{\text{temp}}, 35\ ^{\circ}\text{C}\}$$

Every message that needs to be shared consists of context type and respective value. The importance of the message is calculated based on the mapping of the message and its current context. In above example, the context type is temperature, the possible values of temperature are {10–45}, the selection criteria are the time of temperature recording, i.e. {04:00–23:00} and mapping of all above attributes.

3.3.2 *Mobility*

Another challenge is the mobility. It is the ability of a system that allows the entities to move freely with ease. As the devices are growing in the same proportion as the users are expecting to increase their requirements that applications should satisfy. There are several mobility concerns: the hardware or device mobility, user mobility, linking of the networks.

The device mobility refers to the dynamic configuration of hardware or software in a pervasive computing system. The pervasive system should be able to understand the changes required to handle the device mobility.

In the pervasive computing system, there are several computing environments. The environment could be homogenous or heterogeneous computing environment. The user should be allowed to access the application in the homogenous or heterogeneous or both the environments. The challenge is to seamlessly migrate the application across the computing environment.

For any cases above mentioned (device mobility or user mobility), the most important dimensions or issues are what to move, when to move, where to move and how to move. Mobility refers to the migration of the state of the services from one to another environment. The state of the devices could be a hard state, soft state, etc. In the device or user mobility, the contents stored in the persistent stage and volatile state of service being used by the user must be considered. These things (file, device, service state) are the things that need to be moved seamlessly (what to move).

The dynamic configuration during mobility should consider the environment change or platform change. This 'where to move' concern should be addressed that considers the environment. The characteristics of the environment must be studies well to achieve mobility.

Also, user input (explicit) or the user indications (implicit) choices have to be considered to address 'when to move' dimension during mobility. The explicit input can be the services offered by the application based on the user's location.

During mobility, another dimension is 'how to move'. The state of the user or device should be maintained and should not have any loss to this. Here the approach could be the state storage. Before mobility, the state information should be stored that allows carrying minimum information to carry.

Apart from device and user mobility, another challenge is linking of networks. As we know the pervasive computing involves several networks (wired or wireless or combinations). During mobility, the concern is wired and wireless communication. The short- and long-range communication need to be considered. Also, the speed of communication should be achieved to avoid loss of service or data.

3.3.3 Data Management

Data in the context-aware pervasive system is the context and its information. The context is any information used to define the situation or a person. The varieties of applications have several entities that are communicating with each other. During communication, the asynchronous mechanism is established that help to converge the variety of services, devices, etc. This convergence leads to the context or data generation. The context is data in the context-aware pervasive computing. The data management is a dynamic process which targets to manage the context efficiently.

Sensing and collection of contexts are the first step in the context-aware system. This step includes the exchange of identities of communicating parties. With the help of sensors, the context is sensed. Generally, the sensors are embedded in handheld devices or they are part of the surrounding environment and the issues to understand the different type of sensors, environments, their structure (static or dynamic). Also, the challenge is to sense and gather the context information in quick time as the context is generated in high quantity and in less time.

Further, the challenge is to understand the context and its representation. The context needs to be structured in a standard format since context is shared among common space where different entities are connected. During context management, the storage of contextual data is another challenge the developers should address. In the pervasive system, the entities and their preferences are necessary. Hence, the contextual data and its history need to be stored. This leads to several issues like storage memory size, static/dynamic storage, handling outdated data, security measures, handling missing data.

In terms of interpretation of context, the next challenge is to develop a scheme of converting low-level contextual data into high-level meaningful context. User's intention and mood should be identified from the interpreted context to provide the needed and useful services. The issue is to identify the scope of the based on the context. Understanding user requirements from the interpreted is the challenge. For example, if a person is working on a project for more than 4 h, then possibly he should take a break and grab a coffee. The challenge is to develop a mechanism to identify the need (discovery) and consequent actions (delivery).

3.4 Conclusions

Self-adaptive and self-learning are one of the natural attentive facets of a context-aware computing system. The context-aware computing system should sense the context, the context information and must recognize it type to invoke the appropriate application services. The system should support run-time environment and must handle their behaviour according to the user context and implicit requirements.

Proactiveness and transparency are the key considerations in designing context-aware networks. This chapter provides an opportunity to develop context-aware computing system. The proactive and self-adaptive network will help to handle hybrid applications. The context-aware application or system has many challenges in the future. This chapter has proposed important issues like routing, mobility and data management.

References

1. Barkhuus, L., Dey, A.: Is context-aware computing taking control away from the user? Three levels of interactivity examined. In: International Conference on Ubiquitous Computing, pp. 149–156. Springer, Berlin, Heidelberg, Oct 2003
2. Alegre, U., Augusto, J.C., Clark, T.: Engineering context-aware systems and applications: a survey. J. Syst. Softw. **117**, 55–83 (2016)
3. Loke, S.: Context-aware pervasive systems: architectures for a new breed of applications. CRC Press, Boca Raton (2006)
4. Perera, C., et al.: Context aware computing for the internet of things: a survey. IEEE Commun. Surv. Tutor. **16**(1), 414–454 (2014)
5. Salehie, M., Tahvildari, L.: Self-adaptive software: landscape and research challenges. ACM Trans. Auton. Adapt. Syst. **4**, 1–42 (2009)
6. Brun, Y., Serugendo, G.M., Gacek, C., Giese, H., Kienle, H., Litoiu, M., Muller, H., Pezze, M., Shaw, M.: Engineering self-adaptive systems through feedback loops, In: Software Engineering for Self-Adaptive Systems. Lecture Notes in Computer Science, vol. 5525, pp. 48–70. Springer, Berlin (2009)
7. IBM Corporation: An architectural blueprint for autonomic computing. White Paper, 4th edn. IBM Corporation (2006)
8. Indulska, J., Sutton, P.: Location management in pervasive systems. In: Proceedings of Workshop Conference on the Australasian Information Security ACSW Frontiers, vol. 21, pp. 143–151, Jan 2003
9. Cheng, B.H.C., de Lemos, R., Giese, H., Inverardi, P., Magee, J., Andersson, J., Becker, B., Bencomo, N., Brun, Y., Cukic, B., Serugendo, G.M., Dustdar, S., Finkelstein, A., Gacek, C., Geihs, K., Grassi, V., Karsai, G., Kienle, H.M., Kramer, J., Litoiu, M., Malek, S., Mirandola, R., Muller, H.A., Park, S., Shaw, M., Tichy, M., Tivoli, M., Weyns, D., Whittle, J.: Software engineering for self-adaptive systems: a research roadmap. In: Software Engineering for Self-Adaptive Systems. Lecture Notes in Computer Science, vol. 5525, pp. 1–26. Springer, Berlin (2009)
10. Tennenhouse, D.: Proactive computing. Commun. ACM **43**(5), 43–50 (2000)
11. Zampunieris, D.: Implementation of a proactive learning management system. In: Proceedings of E-learn, pp. 3145–3151. AACE Press (2006)
12. Coronado, S., Zampunieris, D.: Towards a proactive learning management system using early activity detection. In: International Conference on Society for Information Technology & Teacher Education (SITE08), vol. 1, pp. 306–311. AACE Publishing (2008)
13. Dobrican, R., Zampunieris, D.: Supporting collaborative learning inside communities of proactive through proactive computing. In: Proceedings of the 5th Annual International Conference on Education and New Learning Technologies (EDULEARN13), pp. 5824–5833 (2013)
14. Krupitzer, C., Roth, F.M., Van Syckel, S., Schiele, G., Becker, C.: A survey on engineering approaches for self-adaptive systems. Perv. Mobile Comp. J. **17**(B), 184206 (2015)
15. Mansor, A.A., Kadir, W.M.W., Elias, H.: Policy-based approach for dynamic architectural adaptation: a case study on location-based system. In: 5th Malaysian Conference on Software Engineering (MySEC), pp. 171–176, IEEE, Dec 2011
16. Klus, H., Niebuhr, D., Rausch, A.: A component model for dynamic adaptive systems. International Workshop on Engineering of Software Services for Pervasive Environments: In Conjunction with the 6th ESEC/FSE Joint Meeting, pp. 21–28, ACM (2007)
17. Mizouni, R., Serhani, M.A., Benharref, A., Al-Abassi, O.: Towards battery-aware self-adaptive mobile applications. In: Proceedings of the 9th International Conference on Services Computing (SCC), pp. 439–445, IEEE, June 2012
18. Karadimce, A., Bogatinoska, D.C.: Using hybrid mobile applications for adaptive multimedia content delivery. In: Proceedings of the 37th International Convention on Information and Communication Technology, Electronics and Microelectronics (MIPRO), pp. 686–691, May 2014

19. Al-Karaki, J.N., Kamal, A.E.: Routing techniques in wireless sensor networks: a survey. Wirel. Commun. IEEE **11**(6), 6–28 (2004)
20. Sankarasubramaniam, Y., Akan, Ö.B., Akyildiz, I.F.: ESRT: event-to-sink reliable transport in wireless sensor networks. In: Proceedings of the 4th ACM International Symposium on Mobile Ad Hoc Networking & Computing (MobiHoc), pp. 177–188 (2003)
21. Kannan, R., Sarangi, S., Iyengar, S.S.: Sensor-centric energy-constrained reliable query routing for wireless sensor networks. J. Parallel Distrib. Comput. **64**(7), 839–852 (2004)

Chapter 4
Context-Aware Computing and Personalization

Abstract User modelling plays the main role to support an individual user to improve in his/her working experience, learning methods and information access. In the new generation of pervasive and context-aware systems, contextualized and personalized information and services are the driving entities. This chapter explains how context-aware computing is important for personalization. The relationship between machine learning, ontology learning and personalization is also discussed in this chapter. The next part of this chapter focuses on context-aware profiling and profile translation towards context-aware services. The need for clustering in ubiquitous computing and how context-aware clustering will play an important role in the next-generation Internet is discussed in the last part of this chapter.

4.1 Context-Aware Information Retrieval

Due to broadband technology and availability of the Internet widely, Internet computing has been evolved to the Internet of things and eventually to the Internet of everything. Today, a petabyte of data, i.e. 1,024 terabytes, to be exact may probably meet many people's definition of big data. However, in 2025 petabyte will also not qualify as big data. The key elements to contribute to this big data are sensors, radio-frequency identification (RFID) and actuators. As per the statistics of the sensor market, it was $15.27 billion in 2013 and is likely to increase till $29.25 billion by 2020 (CAGR of 9.7%). This growing amount of data causes several challenges when we have to draw specific inference out of this big data. The challenges regarding storage are how we store that much data? How we move that much data? And how we extract, load and transform that much data? However, due to advancement in the semiconductor industry and VLSI technology, space is not the issue. This extensive amount of information requires enhanced search and analytic methods. In the sequel, few challenges are how we explore and analyse that much data? And how we process and get meaningful insights from that much data?

P. N. Mahalle and P. S. Dhotre, *Context-Aware Pervasive Systems and Applications*, Intelligent Systems Reference Library 169,
https://doi.org/10.1007/978-981-32-9952-8_4

4.1.1 State of the Art

Contextual information leverages information retrieval and enriches users with ease of use and personalization. Relevant contextual information and its identification is a key parameter for contextual information retrieval. Massive data uploaded on the cloud lead to information overload and information retrieval (IR) is the only solution, e.g. Google Search Engine. A recommender system like movie recommender is another solution for information overload. Mobile computing and smart home devices help to address context-aware retrieval for information overload. IR is stated as [1], and IR is fundamentally a measurement $\boldsymbol{M}$ for the relevance $\boldsymbol{R}$ of an object $\boldsymbol{O}$ for information need of the user. The need for information is always represented by a query $\boldsymbol{Q}$ and information which is implicit is represented by $\boldsymbol{I}$ and inferred information from $\boldsymbol{I}$ is denoted by $\vdash$.

As stated in Chap. 2 of this book, search in Google by time, search in Google Map by location, searching the hotel by neighbourhood and features are a few examples of context-aware IR. Building smart applications by adapting to the changes in contextual information is referred to as context awareness. As stated by Finkelstein [2], to satisfy user's intentions, more accurate results can be obtained if we carry out a search in line with the information surrounding them. Context-aware IR involves the original whole data, its related data and corresponding structures used in IR. Ubiquitous device, task, data source and geographical location mainly act as a context in IR. There are varied types of context-aware IR, and it varies in the manner these applications interact with the user. Foundation of context-aware computing and applications is nicely presented and discussed in [3].

In context-aware IR, there are two types of applications like interactive applications and proactive applications [4].

1. ***Interactive Applications***

It is a user-driven approach where a user sends a request which also includes context information. The main aim of this request is to retrieve required and relevant resources. Example of this application is which are popular theatres near Pune station? It is important to note that there are no location detectors or sensors deployed nearby and query consists of contexts.

2. ***Proactive Applications***

It is an author-driven approach where every resource is associated with the context. The access to resources is granted to the user if there is a match with the context. Example of this application is to search theatres based on location and time like theatre near Pune station which is multiplex only and having multiple shows after 6 p.m.

Table 4.1 clearly differentiates interactive and proactive applications.

In addition to the classification stated above, context-aware IR also can be personal or shared. In a personal system, IR is tightly coupled with a single user. However, in the shared system, users can be either mutually exclusive (behave like a personal

Table 4.1 Comparison of interactive and proactive applications

Sr. No.	Interactive	Proactive
1	A query is issued by user mandatorily	It is not compulsory that the query is issued by the user
2	User inputs are associated with context information	There is proactive capturing of context
3	It is processed from user to resources	It is processed from resources to users
4	There is no match between the contexts	There is matching between the trigger context and user context

system) or users are interdependent to each other. An example of the shared system is: consider the Department of Computer Engineering. Users are all department staff members and monthly attendance of the students when triggered is sent to the Head of the department.

Issues and Challenges in Context-Aware IR

There are many issues which are specific to IR where the context is associated with accessing information. Few issues are listed below:

1. It is important for an application to keep track of which resource has been processed in the past. This information can be useful to avoid repetition of information, to enhance user awareness and to update about document revision.
2. To improve on the prediction, it is equally important to store previous values and context between inter-retrieval operation.
3. It is also important to keep track of the rate at which the particular resource is being changed so that we can forecast the importance of the resource and optimization can be carried out.
4. Human–computer interaction and usability engineering issues are crucial mainly in mobile applications.

4.1.2 Machine Learning Approach

A machine refers to the computing device with the capabilities of sensing, computing and communication. Learning is the process of acquisition and performance enhancement through experience. Learning also enables the building of memories and behaviours which include abilities, acquaintance, understanding, beliefs and intelligence. Fundamentally, machine learning is a mechanism to automate the formulation of an analytical model which can enable algorithms to learn constantly and optimizing the performance using sample data or past experience. It is really important to understand when would it really require machine learning? Following are a few situations when we require computers to learn:

- The underlined computing problem cannot be solved with the help of available human expertise or skills.
- It is difficult to design an algorithm which is time and space efficient and cannot be a program.
- The problem involves uncertainty and there is a need for a dynamic solution which keeps on changing with time.
- When expertise or skill set in unexplainable.

To apply machine learning to solve the problem, we have to write the program with a certain number of parameters and the appropriate machine learning algorithm will investigate a few parameters which will optimize the desired criteria. In data mining and context-aware IR, machine learning plays an important role in the discovery of desired patterns. As stated earlier, due to ubiquitous computing and advancement in VLSI technology, computing devices are becoming cheaper, powerful and are widely available. As a greater number of devices are connected to the Internet, abundant data is posted on the cloud leading to big data. In the sequel, there is a need for a high-performance IR algorithm. A thorough study is required to design a new algorithm, to decide which algorithm should be applied to solve a given problem which can optimize the performance.

Machine learning algorithms are mainly classified as supervised, unsupervised and reinforcement learning. The selection of a machine learning strategy is based on the domain of the problem and the type of sample data. In ***supervised learning***, a set of inputs along with the corresponding outputs are given, and the objective is to find correct outputs for given test inputs. Classification and regression are examples of supervised learning. In ***unsupervised learning***, a set of inputs are only given, and a set of outputs are not available. The objective is to find some (hidden) structure in a bunch of unstructured data. Density estimation, clustering and dimensionality reduction are examples of unsupervised learning. In ***reinforcement learnin***g, a set of inputs are available from the environment and produce action sequences. It is basically an interaction between a model and the environment.

Machine learning plays a key role in context-aware IR mainly two phases like during drawing inferences from context information and finalizing decisions. Various operations during inference phase are pattern discovery from interaction archival and forming clusters from the user's profile. Decision phase includes user behaviour prediction, task atomization and learning about user preferences. The major goals of learning are association, sorting, recommendation, deletion and creation operation. In literature, machine learning techniques have been effectively applied for context-aware IR. Parametric evaluation of various machine learning algorithms for context-aware IR is presented in Table 4.2.

Table 4.2 Parametric evaluation of various machine learning algorithms

Algorithms	Types of learning	Class	Constraints	Recommended areas
K-Nearest Neighbours	Supervised	Input-based	Does not make any assumption as it is nonparametric	More suitable for the problem which is distance based
Naive Bayes	Supervised	Probabilistic	It works only on the problem where inputs are mutually exclusive	More appropriate for the problem where every class is associated with the probability greater than zero
Bayesian Network	Supervised	Probabilistic and statistical	Computationally expensive and not suitable for a large data set	Suitable for the problems where there are more missing data entries and over-fitting of data can be avoided
Decision Trees	Supervised	Hierarchically organized in the form of Trees	It is more biased about the problem with more levels	More suitable for the problems where data is categorized properly
Fuzzy Logic	Supervised/unsupervised	Fuzzy data	Not suitable for critical application and requirement	More appropriate for approximation problem
Support Vector Machines	Supervised	Decision boundaries	Does not work for the problem where there is no clear classification available	More suitable for binary classifiers
Hidden Markov Model (HMM)	Supervised/Unsupervised	Markov Chain	Only work for the problems where all conditions of HMM holds	More appropriate for temporal data set and stateless information
Neural Network	Supervised	Approximation	Optimal limitations	More suitable binary data set
Clustering	Unsupervised	Cluster-based	Appropriate for all data sets	More appropriate for grouped data set may be in the form of distance
Feature Selection	Unsupervised	Factorization	Appropriate for all data sets	Appropriate for overlapping information

4.1.3 Ontology Learning Approach

With the increasing demands of users and due to economies of scale, data, data set and data structures are becoming more complex. The main goal of ontology is to provide a common and easy understanding of the structure of information to users [5]. Consider an example where there are different websites which contain agriculture information and provide product and fertilizer recommendation services to the farmers. If all these websites provide and publish common ontology of the terminologies and notations they are using, then it will be very easy for farmers and computing agents to retrieve information from these different sources more efficiently. This will eventually lead to effective IR and better user satisfaction. The ontology represents easy and common language or terminology for domain expert who is responsible for sharing expertise and information in a domain. Reusability of the domain knowledge is another objective of ontology where ontologies developed by one group of domain experts can be used by another group of domain experts. In addition to this, to create an ontology which is large in size, several such ontologies can be integrated together to produce bigger ontology. Domain knowledge analysis is equally important if the terminologies are properly defined in the ontology. Reusing existing ontologies or extending them to other application requires formal analysis of the terminologies [6]. In the literature, ontology is defined in many ways. In [7], ontology is defined as representation and description of domain expertise, their respective properties and set of restrictions. The ontology also helps to deduce more accurate contextual information with the help of logic reasoning mechanism [8]. Advantages of using ontology are listed below:

Information can be expressed in a better and effective way with the help of ontologies.

1. Intangible knowledge is represented with more precise representation as compared to other representations.
2. Ontologies are flexible and can be integrated with any rule language.
3. Ontologies are useful to remove ambiguities.

Cost-effective building of ontologies and applying the same to solve the knowledge acquisition problem by developing automatic methods for knowledge extraction about the underlined domain is referred to as ontology learning (OL). OL is an art of building data and knowledge. The source of knowledge (either a scratch or prior knowledge) and the type output used by the learning process are two main aspects of OL. For OL, the input can be in the form of structured data like databases, semi-structured data like dictionaries and unstructured data like document presented in natural language or HTML pages. In the literature, there are many proposals where OL has been applied to be structured as well as semi-structured and the results are discussed for comparison of performance in both the approaches [9]. When OL is applied to structured data, the key issue is to decide the correct part of structural information for knowledge discovery. Databases and knowledge bases are examples of structured data. In E-Commerce and M-Commerce era, the majority of the data is

Table 4.3 Key challenges in OL

Sr. No.	Challenges	Details
1	Heterogeneity	In web computing, data is available in various formats as well in various languages, domains and nature
2	Uncertainty	More uncertainty is associated with poor quality and unstructured data and application of OL to this type of data lead to imperfect results
3	Reasoning	Applying OL to applications where logical inference to be drawn is more challenging
4	Scalability	Due to the Internet of things, more devices are connected to the Internet and resulting in big data and requires more efficient and scalable OL approaches
5	Quality	For obtaining precise, correct and unambiguous results, we need more accurate OL techniques
6	Usability	Knowledge extracted by applying OL should be usable and useful

available in the form of unstructured format, and there are more challenges to apply OL to this unstructured data. All Word and PDF documents, as well as web pages, represent unstructured data, and knowledge extraction from these kinds of sources requires statistical and linguistic techniques and also requires domain knowledge. A statistical technique uses the method frequencies and occurrence of the words in the document, the pattern in which they are appearing as well as the source of the document. Linguistic approaches are based on natural language processing, syntax and semantics. All HTML and XML documents are examples of semi-structured data, and in Web 2.0 computing, the use of this semi-structured data is increased drastically. The quality and usability of the knowledge extracted by applying OL to structured data are better than semi-structured and unstructured data [10]. The reason is, it is easy to define axioms, relations, concepts, synonyms and terms on structured data. Key issues and challenges in OL are listed in Table 4.3.

4.2 Context-Aware Profiling

Profiles play an important role to enable personalized services, and context information is very useful to incorporate in the profile creation process. User profiles help to maximize the value of service to the end-users and will be required when users wish to have services personalized according to their requirements. In pervasive and ubiquitous computing, device profile represents the separate entity which also always follows the users. The profile contains different entities like the identity of an entity for which the profile is created along with several preferences covering context preferences and security, privacy preferences. The user profile represents a user's unique lifestyle. On an overall level, the user profile will consist of at least the following parts:

- The user identity,
- Several types of user preferences,
- Information about common roles undertaken by the user,
- Information about the user's devices, and
- Information about subscribed services.

While building the user profiles, existing templates and wizards must be considered in order to guarantee the dynamic and flexible nature of the user information. It may contain a lot of data and facts such as name, sex, nationality, date and place of birth, address, e-mail address (es) and phone number(s). There must be a record of the device properties or device profiles for the devices that the user possesses. Device profiles would then be used to adapt the service and deliver the content to the user. As stated earlier in the Chap. 1 of this book, there is n number of devices deployed for different purpose and context, and device profiles must be designed by considering the type of application in which they are being used. In the Internet of things, devices deployed are having limited resources in terms of, e.g. processing power or battery life. Based on different attributes of devices, virtual identities can be created [11]. Context-aware profiling enables adaptive and proactive service provision and helps to provide personalized services using user preferences. Context-aware profiling is introduced in 3GPP and is implemented as a virtual home environment use case [12]. XML and other markup languages like SAML and WSDL are used to ensure interoperability because in the case of mobile applications, profiling information is exchanged across multiple boundaries and operators. Context-aware profiling creates more impact on the deployment of service, and its execution as related services based on the context should be adaptive in nature.

4.2.1 Motivation

Mark is a telecom expert and is a sales and marketing professional. His job profile is to travel across the globe for promoting the product and services of his telecom industry. He can access all the information stored on his cloud storage and can do anything, anytime and anywhere kind of computing through his smartphone which is IoT enabled. When he is on one of his business trips, he meets with a very interesting service provision use case which is described below:

On the fly, Mark is receiving real-time videos of the meeting and discussions held at the main office from unauthorized application provider on his large screen Wi-Fi-enabled smartphone while at home. On the fly, when he reached the hotel, being a mobile user, he is parking his vehicle in his hotel parking and the context is changed from a large screen device to a portable 4G device. In view of this, authorized application provider will be notified by context information server about the change of location, the device as well as the type of radio access. In this type of use case, with the help of user's profile, the application will correctly identify the user and as per the context change, there will be a proactive change in the contents from video to text and accordingly, the media converter software agent will successfully execute the conversion of content from one form to another form.

This is one of the scenarios in the context-aware pervasive and applications. In practical deployment and successful operation of such a scenario, user's profiles and their context-aware profiling, the status of the network and system as a whole should be able to meet with large contextual information. In order to drive this kind of use case successfully, the following technological enhancements should be met by underline networks and infrastructure:

- Sensing agent should successfully able to sense the contextual information of the user like location, type of radio access and inform the underlying application about the context change.
- User preferences should be location dependent in order to retrieve application instances as per the context change.
- The pair of preferred geographical areas and set of user preference should be defined in advance to enable application customization.
- For disaster management, alternative approaches should be well documented.

4.2.2 Profile Translation

A profile describes a particular entity like user, device, etc. with some concrete information. To describe the user profile, this information represents user preferences and routine habits. A profile plays an important role in personalization which enriches the use of web services for the user. As the Internet is widely available at cheaper rates, the use of a smartphone is sky-rocketing. Providing a user profile to the service provider is a time-consuming task. In the era of E-Commerce and M-Commerce, Internet transactions are scaled up and for every transaction; the user is required to provide a complete profile. Practically, every time this complete information is not required. However, only specific and relevant user information is required depending on the type of transaction. In existing solutions, user profiles are stored at the service provider's side and one user uses several services and multiple copies of users profile are stored at different places. It is also observed that the user's profiles are changing dynamically with respect to a context like time, space, interest and days, etc. User also needs to memorize these credentials used for profile creation at different places and update the same. This mechanism also results in storage required for profile databases faces the privacy and consistency issues. In view of this, there is a need for proactive profile translation. Context-aware proactive profiling is one of the solutions to avoid this tedious process. In proactive applications, appropriate recommendations are pushed by the system depending on current situations. In profile translation, when the user opens an application on, then proactively profile is generated and according to the context, profile translator will fetch appropriate values from the database.

4.3 Context Management and Personalization

Integration of contextualization and personalization enriches personalized information discovery and delivery and ease of use. Location-based services or network-aware services are the best examples of personalized services where information delivery is based on the user's profiles and logs. This section focuses on the personalization in context-aware pervasive systems and how it helps to provide recommendations to mobile users.

4.3.1 Personalization

Consider the extended user case of Mark. While Mark was roaming in the city, he uses GUIDE VISITOR feature on his smartphone. The device and service discovery module on his smartphone provides detailed information about new things and services like shopping, lucrative discounts on offers, visits to popular local places which are available from different ISPs. At the same time, he also receives a proximity alert from his friend who is also visiting the same city. This is referred to as personalization example which includes an update from online location services and location-based policies. On the fly, he meets with his friend, and they plan for lunch. They decide that they will use his friend Bayu's car, so they can present a temporary insurance contract for using the car. Then, Bayu finalizes a short-term contract with the insurance provider and transfers the required details to Mark. While travel, this car proactively keeps track of Bayu and pays for the tolls as well as parking fee and the same is diverted to Bayu's account instead of Mark account. This is an interesting example of intelligent context management and personalization.

Information and services are accessed from multiple heterogeneous devices, and in this case, the information requires filtering, local adaptation and customization. Information can be adapted as per the needs, interest, objectives and choices with the help of personalization and always contextualization complements personalization. Context-aware user profiles are essential to enrich personalization and enhance the value of services to all users. The design objective of any framework should be to provide fully distributed, mobile, personalized and proactive access to all required services. User exhibits several roles in the various contexts, and there are different user preferences for each role. The sample taxonomy of user profiles based on the role is shown in Fig. 4.1.

This hierarchy of user profiles is extremely useful in deciding the data duplication and helps to enable personalization. Knowledge from the contextual information is used to adapt behaviours of an entity, and this adaptation to the situation enables personalization [13]. Adaptation to situations will to a high degree be determined who the user is and what role the user has, e.g. as a professional journalist the user will require the system to, e.g. use different parameters for discovering services (when using context-aware service discovery), but when the user comes home from work,

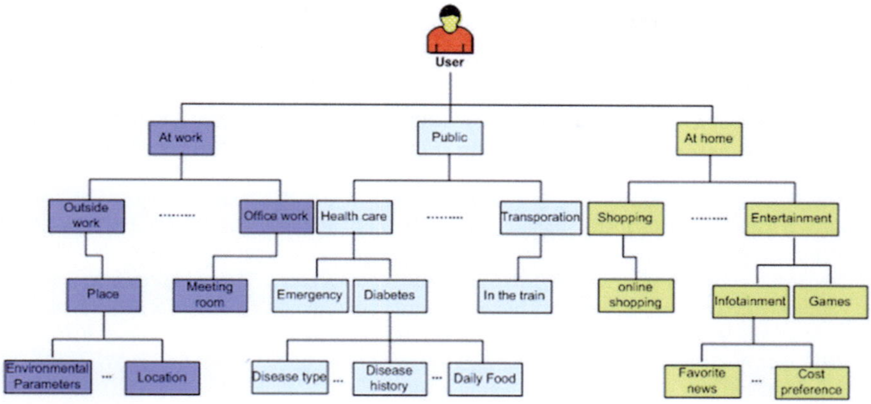

Fig. 4.1 User profiles based on roles

he will no longer have the same (or have different) requirements for the same services he will use.

The benefits of personalization are as follows:

- Personalization helps to proactively provide what they need without an explicit request.
- It also helps to adopt appropriate functionality to interpret user needs from the history of interactions.
- Personalization also helps to enable personalized search.
- It enhances the recommendations generated to satisfy user needs.
- It also provides meaningful information for ranking users and resources.
- Time-saving and ease of use is another potential outcome of personalization.

4.3.2 Recommendations for Mobile Users

In the era of ubiquitous computing, a smartphone is the most popular computing device enabling personalized access to services. However, personalization in smartphone needs several enhancements as compared to the desktop system, and there is a great scope to initiate research in this field. Due to resource constraints like limited memory, battery and computing power, it is difficult to run heavy applications on the smartphone. In addition to this, there are several sensors and services continuously running. In the sequel, there is a need to design a lightweight application to process big data. An application of the context-aware system for an understanding of user's information and eventually timely and relevant recommendations is a need of today. Context-aware search engine and corresponding module play a very important role in generating recommendations for mobile users. Personalized search is one of the outcomes of recommendation where for the given query, results are generated as per the user needs. It combines query and specific context and ranking are generated which is most relevant to the given query and context. Mobile device consists of many resources, and services are running all the time. Tag ranking with respect to the relevant resources so that required resources always appear at the top is referred to as tag recommendation. Consider the following scenario:

> *Mr Mark is planning to attend the International Conference on "Internet of Things, Next Generation Networks and Cloud Computing (ICINC) 2019" that will be held in Mumbai, India. Mark is a very technocrat person; he creates the group "Let us explore ICINC 2019". Mark has added information related to ICINC in this group such that URL of website ICINC, web links of videos related to Mumbai's historical attractions and do's and don'ts while staying in Mumbai. He kept facility to join this group, as his fellow mates/Participants of ICINC-2019 may also join this group. He has tagged information of this group by "ICINC2019", "Conference" and "2019" for information retrieval.*
>
> *Miss Camila, a researcher would like to submit her research article in conference related to her work, hence she was browsing, and she came to know about "Let us explore ICINC 2019" group. She further browses this group as she was interested to publish her article in the conference. However, she learns that the date of the article submission was over, so she tried to look for other conferences where she can submit her article. For information retrieval she used to tag "conference", however, resulted conferences were not related to her interest. Hence, she tried to tag "2019" and again she didn't get the conference she was expecting. To search conference of her interest the search system should combine tag of her interest and year 2019. Furthermore, she is also interested to communicate with an expert in her domain with whom she can discuss her findings and elaborate more on her research.*

Lot of potential work has been done in the field of recommendations for mobile users. Recommendations and personalized services based on the location are presented and discussed in [14, 15]. The use of sensors and GPS is explored in their proposed work. The potential use of a context based on the questions like who are involved in the interaction, when interaction happened and where it happened is presented in [16]. As stated earlier, context management and recommendations based on the ontology model is a widely used method. Lightweight reasoning mechanism based on the graph is presented in [17]. In the proposed pioneer and foundation work [17], reasoning computations are done on the smartphone locally which helps to alleviate security and privacy issues. To design an accurate recommendation system for mobile users, the information needs of users should be clear and precise. User behaviour, preferences and personal nature of the mobile device are important for effective searches and recommendations [18].

4.3.3 Context-Aware Services

With the rapid advancements in mobile computing, development is moving towards proactive mobile systems where applications can be developed using either application-transparent approach or application-aware approach. System software, as well as network software along with all the system parameters, is used in application-transparent approach in order to control several operating conditions like bandwidth, battery level, memory and computing power. However, in application-aware approach, the convergence of system and application software is applied to manage computing and communication resources. Context-aware computing is the next step of application-aware approach where contextual information (like vicinity, location, etc.) and contextual services are used to develop the application. As stated earlier, context-aware applications require contextual information which can be collected from the various devices deployed in the surrounding which has sensing, computing and communication capabilities.

Let us consider the ubiquitous use case to exhibit the example of contextual services. For example, Yashita is visiting Pune, India, for the first time. She has her smartphone, which is equipped with wireless service provided by a service provider providing the services in the Pune vicinity. Due to a busy schedule, Yashita has not made any hotel reservation before coming to Pune. Fortunately, the wireless service provider has Context-aware Personalized Web Information Service (CaPWIS) as a recommender system. CaPWIS considers Yashita's tour plan, location and preferences and recommends the list of hotels which are nearby which matches with her preferences. Yashita then selects the hotel of her choice and initiate online hotel booking. Maximum details in the online registration form are filled proactively from her preferences as the proactive mobile system reads from her profile. Her profile reveals that she always opts for credit card payment, CP plan for a meal, preference for a non-smoking room, etc.

Scenario stated above explains that heterogeneous contextual information is required to make application responsive and personalized. There are several middleware architectures which are used to develop context-aware applications. The objectives of using middleware architecture are to provide contextual services which can be directly used to develop applications [19, 20].

1. ***Context subscription and delivery***: this is a service to which all applications are subscribed. This service is also responsible for sending context event back to the application.
2. ***Context query***: this service helps the application to query for the current context.
3. ***Context transformation***: this service basically does the job of the presentation layer. It transforms contextual information represented in a low level to a high level so that it can be easily understood by the application.
4. ***Context synthesis***: This service is accountable for a fusion of multisources contextual information.
5. ***Discovery and management***: service discovery and management of sensors and connected software are done by this service.

In addition to the services mentioned above, leading industry partners also presented other contextual services for the delivery of better customer experience and analytics, more sale and lower cost. These contextual and user-driven services are:

1. ***Pre-emptive customer service***: many times, the customer has to repeat the same steps which have been already done to the agent, which become more frustrating to the customers. This service is responsible to present customers past behaviour to the agent.

2. ***Services within services***: This service is responsible for collaborating and sharing data across various services in order to provide better-converged service. For example, the user can book a cab from Facebook messenger without switching to cab app.
3. ***Paying without paying***: this service helps to initiate on the fly payment based on the location as contextual information.

4.4 Context-Aware Clustering

As stated in the previous chapters of this book, context plays a crucial role in every part of the Internet computing. The previous decade has witnessed major advancements in operating mechanisms in the field of communication technologies. Simultaneously, major stakeholders in the communication sector have made huge investments. Today, we are functioning in anything, anytime, anywhere paradigm, and the credit of this goes to broadband technology which is widely available. This has led to an exponential decrease in the cost of Internet exponentially enabling smartphone users to carry the Internet in their smartphone. There are more devices which are being created with Wi-Fi capabilities, and in the sequel, numerous heterogeneous devices are connected to the Internet to enable resource sharing and communication with other participating nodes in the network. In today's pervasive system, all sets of operations are driven by the context in order to bring the proactivity. There are a variety of participating nodes like laptops, PCs and smartphones.

4.4.1 Motivation

In many metropolitan cities, smartphone users travel via public transport. Fingered devices use one or more interfaces such as 802.11, GSM or Bluetooth. Such interfaces enable a variety of services to be shared among the masses via their personal handheld devices. The GPS receivers record the location, and time information broadcasts the information to other devices. There is a need for sharing and downloading software components, new applications and functionalities from peers. Consider the scenario where he/she boards a busy bus that was scheduled on his/her personal digital assistant. In transit, the passenger requests for location information to the nearby devices by updating his/her travel panel. The neighbouring devices receive the request and respond to the request on the fly. After receiving a response from the co-located devices, suitable actions can be taken to respond to the above-stated scenario. This use case clearly explains that the context plays a very important role in communication which helps to uniquely differentiate devices and services over the network.

As mentioned earlier, each network has its own pros and cons. New mechanisms, policies and algorithms are required in order to address the issues pertaining to the context-aware applications in the ubiquitous and pervasive communication networks. Context-aware clustering mechanisms are required in order to build an efficient context-aware network. Clustering refers to the process of creation of logical groups by identifying related patterns or on the basis of information. The resultant groups are thereby called clusters. Homogeneous and heterogeneous networks help in improved utilization of resources which can still be improved with the help of better clustering algorithms. In this view, context-aware clustering mechanisms are required in order to improve the scalability over a wide range of networks. The proposed solution must guarantee the efficient management of the cluster structure, thereby enabling smooth network scaling and uninterrupted usage.

4.4.2 State of the Art

As we are aware, in a wireless sensor network, sensor nodes are deployed across regions of interest in order to collect information about the events of interest. In [21], the authors propose a mobility-based clustering protocol for mobile nodes of WSN. Here, the participating sensor nodes choose one of the nodes in the network as the cluster head based on the nodes' residual energy and mobility. The objective of the sensor nodes except for the cluster head node is to stabilize their links with that of the cluster head node. The proposed protocol uses the TDMA-based data transmission policy which provides each node with the time to send out its data in ascending order. The proposed work helps in improving the delay awareness of WSN. Simulation results prove that the proposed work helps reduce the packet loss up to 25% in contrast to that of the cluster-based mechanisms that are portrayed better performance in an adaptive mobile environment. The energy efficiency of the wireless sensor network poses a major challenge to many researchers. In [22], the authors propose a distributed energy-efficient algorithm that understands the cluster size based on the hop distance to the data sink while recommending approximation equalization of the node's lifetime and reduced energy consumption levels. A simple multihop energy-efficient protocol has been established in order to gauge its effectiveness over efficient clustering and end to end energy consumptions. The proposed work increases the lifetime of the sensor networks. The outcome shown in the paper justifies that the proposed algorithm improves the lifetime and energy equalization compared to two well-matured algorithms HEED and UCR.

It is crucial that the QoS parameters are met in any real-time scenario. Energy efficiency is one such key parameter which varies with time and has an effect on system performance. In [23], the authors propose a protocol takes into consideration the energy constraints of WSN and utilizes a genetic algorithm to calculate the fitness function based on two important parameters, namely distance and energy. The probabilistic searching approach is utilized in the genetic algorithm. A comparative study of the results shows that the proposed protocol works better in comparison to that of the LEACH protocol. The selection of the cluster head can be done on the basis of the fitness function. In the attempt to find the best path, the cluster head of one cluster sends information to the cluster head of the neighbouring cluster. This approach suits real-time scenarios only to a certain some extent. Many popular clustering techniques have been used in real-time applications to categorize numerical data. In [24], the authors propose a mechanism that enables the clustering of categorized data in the course of understanding the context of information. The impact of the proposed methodology is that it can be used in the real-time scenario. The proposed work comes as one of its kind in the combination of text processing and machine learning algorithms. This forms the basic building block of a context-aware approach. To decrease the time complexity of the algorithm, the word embedding is stored in a separate file. The testing of the proposed algorithm has been successfully completed on the zoo data set and the flag data set. The authors have used the traditional k-means clustering algorithm. The proposed work is interactive in nature, i.e. the accuracy of results improves with self-supervision and information generated during one iteration work as feedback per the second iteration.

VANET stands for vehicular ad hoc that has properties such as fast-moving nodes, dynamic topology and node movement. In [25], the authors have emphasized specific applications and collaborations with a game theory for VANET. The proposed work relies on the creation of the cluster, once the convoys temporarily rise in traffic dynamics. In the initial phase, the utility function is applied, neglecting the significance of a node in the network, an ordinary node, cluster node of cluster head or tree node. The authors have proved that the game theory mechanism performs better in real-time than the selfish node approach.

The allotment of sensor nodes over a huge area with random deployment approach is a very simple and cost-efficient method. However, it comes with its own challenges. This gives rise to randomness on densities of a specific location. In [26], the authors propose a globally distributed clustering technique. In this proposed methodology, each node is furnished with some global information about the network to help in the finalization of its own significance in the cluster locally. This assumption has been made by the base station to carry on communication and topological activities. The authors assert that the proposed methodology has improved results than the conventional clustering algorithm, i.e. low-energy consumption and improved lifetime of the networks. In the past two decades, social media has played a vital role in connecting the whole world by sustaining social relationships online. It is interesting to analyse how the information about the clusters is made based on social relations. This social relationship emphasizes the idea of context awareness based on the user's behaviour. In [27], the author proposes a context-aware mobile P2P network framework. It further provides a clustering model and algorithm based on locations of the user and the environment in which the user operates. The integration of all these factors brings a new context-aware cluster of the P2P social network. The proposed method is seen to improve the response time, load balancing and adaptive ability. As far as the practical implementation of the proposed work is concerned, the results may vary based on the dynamic nature of clustering algorithms.

The authors in [28] propose a centralized cluster-based protocol which makes use of a harmony search algorithm. The idea is to decrease the distance between nodes and cluster head, thereby leading to optimization in the energy distribution of WSN. The experimentation on the proposed protocol has been executed over an indoor scenario such as an ambient temperature for fire detection. The results of the proposed work are compared with the existing well-established protocols like LEACH-C and FCM. It also illustrates that the proposed protocol can be used in real time for safety and monitoring applications. The results concludes that the proposed work not only provides the best fitness value but also lessens the computation time by 10 ms in comparison with that of Fem. The proposed framework can be used for building effective clustering protocols.

Table 4.4 depicts the gap analysis of the comprehensive literature survey made in the above section. Here, we consider five context parameters like location, device proximity, environment, device type and RSSI. All these parameters help in realizing maximum context information for intelligent decision-making to the devices. The researchers use different approaches and algorithm to get the best results. Every author has used a different approach for different networks as every network has different requirements and different underlying architecture. It is evident that each clustering algorithm solves the problem for the specific network. Hence, a generic context-aware clustering algorithm that works on every network is required. The main objective of a clustering algorithm is to achieve stability, low maintenance overhead and energy efficiency. Each algorithm works on achieving greater stability with varying parameters which function as inputs to the algorithm in use. The variable parameters are as follows: a mechanism for the selection of the cluster head, the varying hop distance between nodes, the number of participating nodes in the individual cluster and energy-efficient approaches.

These clustering algorithms attain significant performance levels, but the dynamic nature and mobility of the nodes pose a greater challenge towards realizing the objective. A lot more research is needed in this area.

4.4.3 Context and Clustering

Information about specific attributes is needed for the formation of clusters. This information helps in the classification of existing data into groups, and such information is called context. The contextual information smartens the system which in turn leads to an efficient decision and prediction making. The context of a network is determined by its parameters like nodes (computing devices) and links (the communication medium). A novel clustering mechanism can be built by using contextual information. In this chapter, a novel generic framework is proposed which enables heterogeneous networking environments to achieve the properties of efficient service delivery and scalability in the pervasive and ubiquitous network. To implement context-aware clustering, context attributes which extract contextual information need to be defined. As explained in Chap. 2 of this book, parameters like device

Table 4.4 Gap analysis

Existing system	Parameters				
	Location	Device proximity	Environment	Device type	RSSI
Mobility-based clustering protocol for WSN with mobile nodes	Yes	No	Yes	No	No
Genetic algorithm-based clustering approach for WSN to optimize routing techniques	Yes	Yes	No	No	No
Context-aware clustering using glove and *k*-mean	No	No	Yes	No	No
QOS-aware and heterogeneously clustered routing protocol for wireless sensor network	Yes	No	No	No	No
Context-aware clustering in VANETS: a game theoretic perspective	Yes	Yes	No	No	No
Global distributed clustering technique for randomly deployed WSN	Yes	No	No	No	Yes
A context-aware member clustering algorithm based on ant colony and genetic optimization for P2P mobile	Yes	Yes	Yes	No	No
Efficient and secure routing protocol for wireless sensors through SNR based dynamic clustering mechanism	Yes	Yes	Yes	No	No
Real-time implementation of a harmony search algorithm-based clustering protocol for energy-efficient wireless sensor network	No	No	Yes	No	No

type, location, RSSI, device proximity and the environment have been considered. A fuzzy approach is used for context management which is divided into four phases:

1. Defining the context for the computing devices,
2. Defining the parameters for each context,
3. Generating fuzzy rules and their validation,
4. Calculating all the context parameter crisp values using the fuzzy approach.

4.4.4 Context-Aware Clustering

The process starts with receiving the input in the form of context parameters as shown in Chap. 2. The cluster types determine the combination of context parameters that will be received as input. Clusters are formed based on the type of context received. Identification of contextual parameters is a crucial task. Fuzzy approach for context-aware clustering takes into account the contextual information parameters, i.e. location, device proximity, environment, device type and RSSI. The location gives values based on the physical existence of the device. In order to achieve higher accuracy, technologies like the GPS, RFID, Wi-Fi or a combination of two devices are used. The ability of the device to communicate with other devices is termed ad device proximity. The device type provides information on the type of the device identified. RSSI defines how effectively the device connects to an access point in the wireless environment, which determines the signal strength of the device. Hence, context can be defined as:

$$\text{Context} = \{\text{location, device proximity, environment, device type, RSSI}\}$$

Every parameter defining the context is designated with three linguistic variables and depending on the fuzzy rules formed the respective device classification will be concluded. Variables assigned to every context parameter are as follows:

Location	{Close, Near, Away}
Device Proximity	{Excessive, Moderate, Shallow}
Environment	{WLAN, WiMAX, 3G/4G}
Device Type	{WSN, IP, RFID}
RSSI	{High, Medium, Low}.

In the fuzzy approach, the fuzzy function has to be developed with the context information and attributes in order to yield the output of the model. According to the fuzzy rule creation strategy, a maximum of 243 combinations of the available variables is developed. However, only a few combinations stand valid in order to represent the structure of the rule. Selection of the correct combination requires rigorous analysis. For example, a combination of the wireless sensor network as a device type and 3G/4G as the environment does not stand valid as a WSN device cannot be used with the 3G/4G network. Valid and efficient fuzzy rules to arrive at an accurate range can be realized using MATLAB fuzzy toolbox. Fuzzy rule produces an appropriate classification of devices like PC, laptop, palmtops, smartphones, tablets and radio-frequency identifier (RFID) devices, which is then used as an input (context) for context-aware clustering. This fuzzy approach for context-aware clustering is depicted in Fig. 4.2.

Clustering algorithm presented in Fig. 4.2 works as follows:

Clustering algorithms take fuzzy rules as input. Its functions are divided into three phases as follows:

1. ***Context Definition***

The context definition feeds the basic data required for building intelligence into the system. Various techniques are used to define the context. The set of parameters that extort maximum contextual information are used as inputs to the fuzzy approach to realize the context-aware clustering. As mentioned earlier, the context parameters that are considered in the proposed work are location, device proximity, environment, device type and RSSI.

2. ***Context Detection***

Context detection is also an important phase. The context parameter used by the device or network helps in proper identification and delivery of service. The devices' contextual information can be evaluated using the definition of context. Here, leverage the principle of permutation/combination. Consider a scenario, where m variables are mapped to n values, i.e. parameters (A_1, A_2, …, A_n) take values from the set (1, 2, …, n). This gives rise to a total of $m * n$ possibilities. In the definition presented above, there are five parameters that provide contextual information and each parameter can be further defined by a set of variables. According to the above permutation/combination principle, we get the following context rules:

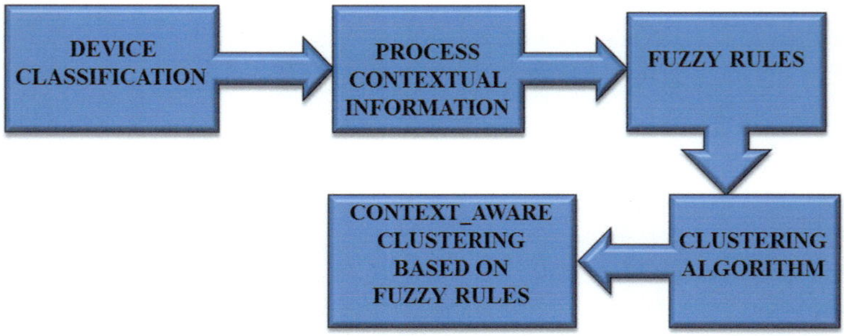

Fig. 4.2 Fuzzy approach for context-aware clustering

Context rules are the number of possible context definitions that exist for a device.

m no. of context parameter.
n no. of values associated with the individual parameter.

The number of contextual rules $= m * n$.
Examples of RULE Definition are as follows:

$Rule_0$ AWAY, DP_1, WLAN, WSN, LOW
$Rule_1$ CLOSE, DP_3, WMAX, RFID, MEDIUM
$Rule_2$ NEAR, DP_1, WLAN, WSN, HIGH
$Rule_3$ NEAR, DP_3, 3G, IP, LOW
$Rule_4$ AWAY, DP_2, WLAN, IP, MEDIUM
… .
$Rule_{243}$ CLOSE, DP_1, WLAN, RFID, HIGH.

Every context definition is mapped to its own context rule. The context definition enables detection of context in the system. The context definition and its associated context rule are picked up in order to apply the context detection policy. This fuzzy approach for context-aware clustering leverages the agglomerative hierarchical clustering method. It improves the scalability and efficiency of the network. This process involved in the clustering mechanism is presented in Fig. 4.3.

The complete process involved in the proposed system is divided into four layers. The process commences with the gathering of contextual information defined by five parameters at each stage. The latest cluster pair introduced integrates well with the older pair of the cluster considering the cluster proximity.

Level 1 cluster (parameterized): in this phase, the contextual information is gathered to define the context. The clusters are primarily formed based on the device type and environment.
Level 2 clusters (application): in this phase, the real cluster formation is done based on user application using the combination of two or more contexts.

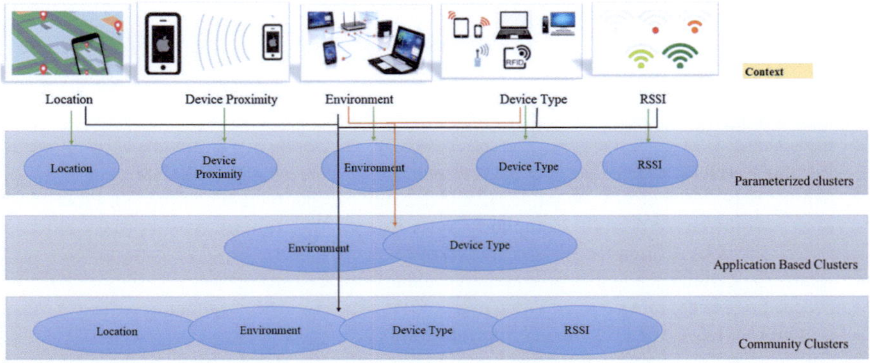

Fig. 4.3 Hierarchical clustering method

Level 3 clusters (community): in this phase, clustering is done using specific parameters like environment, device type, location and RSSI of context. At this level categorization of services for the community, users are done.
Level 4 clusters (improved): this is the final stage where all the clusters are formed by clubbing multiple parameters, depending upon the network type. As mentioned earlier, clusters are formed by considering user needs as every network has its own set of requirement.

4.5 Conclusions

Context-aware computing and personalization is a need of today due to the increasing use of the Internet and smartphone. This chapter has given us an overview of context-aware information retrieval, state of the art as well as issues and challenges in context-aware information retrieval. Machine learning and ontology learning for information retrieval are presented and discussed in the next part of this chapter. Parametric evaluation of various machine learning algorithms and key challenges in ontology learning approach is also discussed in this part. Furthermore, this chapter presents an overview of context-aware profiling and the importance of profile translation in pervasive applications.

This chapter also presents the relationship between context management and personalization and overview of profile creation based on the roles. Recommendations for mobile users and along with the use case are explained in this part of the chapter. Context-aware services and context-aware clustering, as well as motivation with the practical use case, are presented and discussed in the last part of this chapter. Comprehensive literature survey and gap analysis are also discussed in this chapter. Finally, this chapter concluded with the proposed clustering algorithm and its mechanism.

References

1. Dominic, S.: The Modern Algebra of Information Retrieval. The Information Retrieval Series. Springer, Berlin (2008)
2. Finkelstein, L., Gabrilovich, E., Matias, Y., Rivlin, E., Solan, Z., Wolfman, G., Ruppin, E.: Placing search in context: the concept revisited. In: WWW, pp. 406–414 (2001)
3. Schilit, W.N., Adams, N.I., Want, R.: Context-aware computing applications. In: Proceedings of the Workshop on Mobile Computing Systems and Applications, Santa Cruz, California, pp. 85–90. IEEE Computer Society Press (1994)
4. Jones, G.: Challenges and opportunities of context-aware information access. In: UDM'05: Proceedings of the International Workshop on Ubiquitous Data Management, pp. 53–62, Washington, DC, USA. IEEE Computer Society (2005)
5. Gruber, T.R.: A translation approach to portable ontology specification. Knowl. Acquis. **5**, 199–220 (1993)

6. McGuinness, D.L., Fikes, R., Rice, J., Wilder, S.: An environment for merging and testing large ontologies. In: Cohn, A.G., Giunchiglia, F., Selman, B. (eds.) Proceedings of the Seventh International Conference on Principles of Knowledge Representation and Reasoning (KR2000). Morgan Kaufmann Publishers, San Francisco, CA (2000)
7. G'abor, N.: Ontology development. In: Studer, R., Grimm, S., Abecker, A. (eds.) Semantic Web Services. Springer, Berlin, Heidelberg (2007)
8. Noy, N.F., McGuinness, D.L.: Ontology Development 101: A Guide to Creating Your First Ontology, 94305. Stanford University, Stanford, CA
9. Wang, X.H., Zhang, D.Q., Gu, T., Pung, H.K.: Ontology based context modelling and reasoning using OWL. In: Proceedings of 2nd IEEE Conference on Pervasive Computing and Communications (PerCom 2004), Workshop on Context Modeling and Reasoning, pp. 18–22, Orlando, Florida, Mar 2004. IEEE Computer Society Press (2004)
10. Lehmann, J., Hitzler, P.: A refinement operator based learning algorithm for the ALC description logic. In: Proceedings of the 17th International Conference on Inductive Logic Programming (ILP). Springer, Berlin (2007)
11. Dellschaft, K.: Measuring the similarity of concept hierarchies and its influence on the evaluation of learning procedures. Master's thesis, Universität Koblenz Landau, Campus Koblenz, Fachbereich 4 Informatik, Institut für Computervisualisitk (2005)
12. Sarma, A., Girao, J.: Identities in the future internet of things. Wirel. Pers. Commun. **49**, 353–363 (2009). © Springer Science+Business Media, LLC. 200 Tuesday, 07 Apr 2009
13. GPP TS 23.127: Service aspects; The Virtual Home Environment
14. Olsen, R.L., Nickelsen, A., Nielsen, J., Schwefel, H.P., Bauer, M.: Experimental analysis of the influence of context awareness on service discovery in PNs. In: Proceedings of IST Summit 2006, Mykonos, Greece (2006)
15. Sun, X., May, A.: The role of spatial contextual factors in mobile personalization at large sports events. Pers. Ubiquitous Comput. **13**, 293–302 (2009). https://doi.org/10.1007/s00779-008-0203-6
16. Suh, Y., Shin, C., Woo, W.: A mobile phone guide: spatial, personal, and social experience for cultural heritage. IEEE Trans. Consum. Electron. **55**(4), 2356–2364 (2009). https://doi.org/10.1109/TCE.2009.5373810
17. Jang, S., Woo, W.: Ubi-UCAM: a unified context-aware application model. In: Proceedings of the 4th International and Interdisciplinary Conference on Modeling and Using Context, CONTEXT'03, pp. 178–189. Springer, Berlin, Heidelberg (2003)
18. Schmidtke, H.R., Woo, W.: Towards ontology-based formal verification methods for context aware systems. In: Tokuda, H., Beigl, M., Brush, A., Friday, A., Tobe, Y. (eds.) Pervasive 2009, pp. 309–326. Springer, Berlin (2009)
19. Church, K., Smyth, B., Bradley, K., Cotter, P.: A large scale study of European mobile search behaviour. In: Proceedings of the 10th International Conference on Human Computer Interaction
20. Dey, A.K., Sabler, D., Abowd, G.D.: A contextual framework and a toolkit for supporting the rapid prototyping of context-aware applications. Hum. Comput. Interact. (HCI) J. **16**(2–4), 97 (2001)
21. Abowd, G., Dey, A., Brown, P., Davies, N., Smith, M., Steggles, P.: Towards a better understanding of context and context-awareness. In: Handheld and Ubiquitous Computing, pp. 304–307. Springer, Berlin/Heidelberg (1999)
22. Wei, D., Jin, Y., Vural, S., Moessner, K., Tafazolli, R.: An energy-efficient clustering solution for wireless sensor networks. IEEE Trans. Wirel. Commun. **10**(11), 3973–3983 (2011)
23. Nayak, P., Vathasavai, B.: Genetic algorithm based clustering approach for wireless sensor network to optimize routing techniques. In: 2017 7th International Conference on Cloud Computing, Data Science & Engineering-Confluence, pp. 373–380, Jan 2017
24. Juneja, P., Jain, H., Deshmukh, T., Somani, S., Tripathy, B.K.: Context aware clustering using glove and K-means. Int. J. Softw. Eng. Appl. **8**, 21–38 (2017)
25. Chiti, F., Fantacci, R., Dei, E., Han, Z.: Context aware clustering in VANETs: a game theoretic perspective. In: 2015 IEEE International Conference on Communications (ICC), pp. 6584–6588, IEEE, June 2015

26. Abdellatief, W., Youness, O., Abdelkader, H., Hadhoud, M.: Global distributed clustering technique for randomly deployed wireless sensor networks. In: 2016 12th International Computer Engineering Conference (ICENCO), pp. 8–13, IEEE, Dec 2016
27. Cao, H.H., Zhang, Y.M.: A context-aware member clustering algorithm based on ant colony and genetic optimization for P2P mobile social network. In: Applied Mechanics and Materials, vol. 347, pp. 2458–2462. Trans Tech Publications (2013)
28. Hoang, D.C., Yadav, P., Kumar, R., Panda, S.K.: Real-time implementation of a harmony search algorithm-based clustering protocol for energy-efficient wireless sensor networks. IEEE Trans. Industr. Inf. **10**(1), 774–783 (2014)

Chapter 5
Architecture for Context-Aware Systems

Abstract The ubiquitous computing devices, smartphones and applications are being used beyond traditional desktops. Diverse environment, distributed architecture and enhancement in the behaviour are the main challenges towards the emerging ubiquitous computing era. This primarily focuses on the design and development of context-aware applications. This chapter presents architectural requirements and various approaches to develop applications using context. This chapter also presents the standard reference model and application platform, design and development of context-aware algorithms along with issues and challenges. Last part of this chapter presents the importance of machine learning to develop context-aware applications and finally concludes with the case studies.

5.1 Introduction

5.1.1 Architectural Requirements and Approaches

Context awareness is a very important feature of pervasive or ubiquitous computing. The vision of context-aware pervasive computing is developed from a variety of technologies, and it ranges from desktop applications to mobile application. In order to develop a smarter system, there is a need to characterize the context-aware system. The characteristic or properties of contextual information vary from one scenario to another, application to application and system to system. In order to integrate context awareness in pervasive computing, there is a need to look into context-aware architecture for pervasive computing. To design and develop a context-rich system, we need to focus on how the participating devices or nodes perceive each scenario, entity and observations. There are various architectural requirements of the context-aware system. Utilization of context awareness in pervasive computing enables a variety of technologies which improve the comfort level of everyone. In order to improve the performance of the context-aware system, every functionality should be divided into layered style so that in the large system, each component can perform its task independently. The minimum architectural requirements are listed below:

P. N. Mahalle and P. S. Dhotre, *Context-Aware Pervasive Systems and Applications*, Intelligent Systems Reference Library 169,
https://doi.org/10.1007/978-981-32-9952-8_5

1. **Scalability**: The pervasive system consists of a large variety of computing and sensing devices. In a real-time scenario, some devices go down and sometimes we need to add devices in the system so it required to make provision for addition or deletion of devices or things. At the same time, new functionality should able to add without making many changes in the existing scenario. The goal of scalable architecture can be achieved if we design layered architecture. For designing scalable context-aware application, we have three approaches, i.e. horizontal, vertical and hybrid approaches. The vertical approach will be used when the capacity of the system will not be matched to the required capacity. For example, in some cases, we need to enlarge data processing capability, so we need to increase the capability of the CPU. If we are unable to match a number of resources in the same scenario, we need to opt for horizontal scaling. The hybrid approach is more suitable for large systems. In real-time scenarios, the variance of system components are performing their assigned tasks so it is possible that some tasks are needed to perform sequentially, and some of them parallel so it is required to use both approaches wisely by considering a system requirement. It is the responsibility of the system designer to use, which approach at what pace. Concurrency and parallelization are key factors that decide the implementation of above-mentioned approaches in real-time scenarios.
2. **Secure**: The wide variety of applications and their deployment in various domains make a huge impact on society. These pervasive applications bring comfort and security risk as well. In any smart application, many things happen without human intervention, but in such a system, we cannot anticipate the side effects. We are using various location-aware applications, and we are providing location information to customized services. We are using such services but at the cost of giving us personal information. In such scenarios, protecting location privacy is very difficult. In pervasive computing application security professionals play a key role because they need to keep eye on each scenario. They need to think of how many possible ways through the things which could go wrong. In pervasive computing, the devices are continuously joining and leaving the network and form a new ad hoc network from time to time, so we cannot rely on traditional security model. The security spectrum of pervasive computing consists of a very large spectrum, including authentication, authorization, access control, trust management and privacy.
3. **Abstraction**: In a real-time scenario, it is necessary to hide low-level sensing information. The responsibility of system designer is to characterize adoption opportunities by collecting data from a variety of heterogeneous sensors which are networked and continuously sensing, environmental parameters like temperature, humidity, etc., and for data aggregation, we need a software API that collects data from Internet sources or the data generated by wearable devices. This contextual data is large in volume and it is continuously updated and multivariate in nature. It is very difficult to manage the complexity of such data, so we can use data abstraction methods and so developers can easily deal with sensor data.
4. **Adoption of Resource Constraint Devices**: Pervasive computing consists of a wide variety of devices and software agents. Some of these are resource constraint

agents and devices. The technological advancements help to minimize constraints of the resource by offering a memory, computing power and improved sense of accuracy. This helps to satisfy the resource constraint objects. It is much sure that this technological advancement is able to understand contextual information by doing enrichment in software agents and devices. Consider an example, when multiple sensors sending sensed information to its middleware node. The task of the middleware node is to aggregate data, preprocess that data and by performing some data analytics operation take only required decision locally. So, middleware performs multiple tasks simultaneously. It may require that middleware requires more resources due to which middleware unable to develop a complete understanding of contextual information. The probable solution of such problem is sharing knowledge among all other software agents or devices. This process helps to provide a number of resources required to develop or understand contextual information. So, this knowledge sharing process helps to integrate context information into the system very fast; at the same time, we can develop a smart application easily [1].

5.1.2 *Reference Model*

There are various reference models for context-aware systems proposed in the literature depending on the underlined application. However, a more general model is presented in Fig. 5.1.

The physical layer consists of sensors, actuators and other physical devices. The new ideas of these devices are continuously invented. The sensor devices sense the environmental parameter and send that data to the above layer in the architecture.

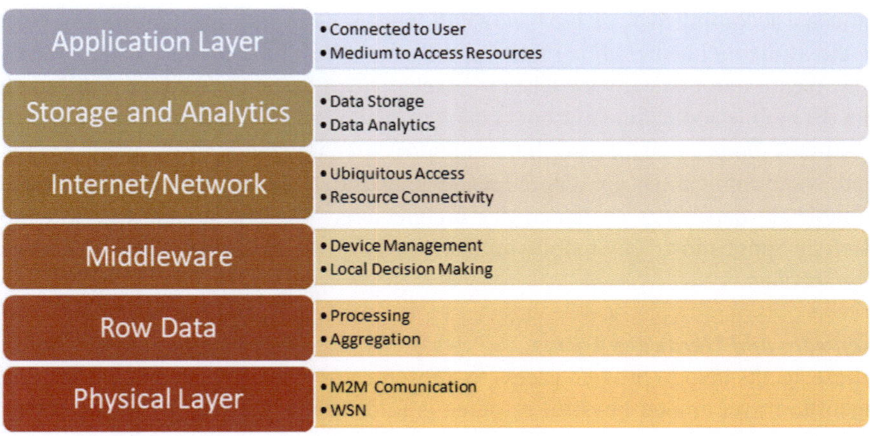

Fig. 5.1 Abstract layered architecture

There are a variety of physical sensors are present like temperature, humidity, camera, microphone, GPS, etc., and there are virtual sensors also available. If we look from the software perspective, we can say that we are using virtual sensors in day-to-day life like exploring electronic calendar, online ticket booking, utilization of devices and bandwidth. Logical sensors integrates physical and logical information like cookies, browsing history from files, etc.

When a variety of sensors of physical layer generate data in a variety of formats, we can say it is raw data because it consists of noise and dirtiness. The qualification of sensors provides quality data. The data aggregation node supplies data to middleware. The responsibility of middleware is to preprocess the data which is provided by the layer below. In a real-time scenario, we need to preprocess data, i.e. removing noise and dirtiness of the data. The improved qualities of data help for better prediction and efficient decision-making. In a real-time scenario, for a certain event, middleware needs to take a decision without forwarding data to the next layer. The middleware also consists of business logic so that middleware is able to take a decision. The middleware supports adaption of the various devices, software agents, mobile agents and auto-discovery of services. The characteristics of middleware are listed below:

Adaption

The exponential growth of mobile devices is observed in the last decade. Most of the users carry at least one mobile device. As far as computing is a concern, the gap is becoming narrower between the mobile user and traditional devices. The applications are hungry about the computation, storage and communication capability. At the same time, we face the technological differences due to advancement and infrastructure for technology deployment. The application developers of the mobile device put their major efforts to make computation comfort for the underlying device and traditional device become so narrow. The application must adopt the behaviour of traditional devices and optimize resource utilization. The quality of service must be adjusted to achieve a moderate level of performance, e.g. the quality of video and audio must be adjusted according to available bandwidth. The user can view the high-quality video if the available bandwidth is high. On the other side, if the quality of the video is low, then it automatically switches to pixeled video based on available bandwidth, but the mobile application must not maintain the lowest quality in the case of opulent resources. Some application optimizes the resource utilization itself, and in another end, some applications are ignored resource optimization and such applications are completely dependent on underlying architecture. We need middleware to adopt such an application. This mobile middleware helps to adapt and manage real-time applications.

Software and Hardware Agents

These agents help to provide access to a resource over the Internet. The task of the agent to travel around physical resource is to distribute the code to the associated entity and collect accessible information from resources or sometimes it carries computational information and returns all these information to central node or server, and after that, it will continuously migrate in the network to collect information. These agents work very intelligently and keep track of information. Consider university paper collection use case, where a software agent collects the set of question from all professors who are investing their efforts for the paper setting process. So, software agent travels node by node and collects an available set of question from the professor's computer. Sometimes these agents face some critical problem like connection breakdown. In such condition, also agents hold the information as it is and when connection establishes, agents forward the information to the base server. This scenario indicates that these agents also support connection breakdown. Consider the above example with the objective of setting up question paper for university exam so there are various paper setters who present different geographical location. These paper setters prepare question paper and keep at a specific location of their laptop or computer. When software agent visits every node, it will collect a set of question and prepare the set of questions which are in the actual question paper. At the same time, these agents try to cover dynamic aspects of the pervasive system. Here we can argue over these points that we can also perform the same task with the traditional methodology by modifying code at servers. Yes, we can do the same things with traditional methods also.

Auto-discovery of Services

This feature in the devices gives another dimension to existing to the client–server paradigm. Now, the interaction becomes more dynamic in client–server communication. Auto-discovery of services is suitable for both wired and wireless networks. As far as mobile computing concern, this feature helps a lot in real-time scenarios because this feature allows mobile devices to get information about services available at that location automatically. Suppose John is technophile and handling marketing department for his company, he has ubiquitous access to the company's resources and information through his latest devices. One of John's business trip is as follows:

When John enters at the airport, he gets all the information about whatever services available at the airport, e.g. guided map of the airport, the location of the cafeteria and a variety of services at the airport.

Then John chooses to subscribe to service that informs about current waiting time, and at the same time, he got a message that to due technical problem the flight is late. So airline provides e-vouchers of lunch.

John's smart device identifies available network at the airport to access office information and services, so he started to work till he meets his flight.

When he reached the destination, he completed his office work and went for shopping. When he was travelling through the market, he got various notifications regarding the offers on those products that he had kept in his wish list.

5.1.3 Context-Aware Application Platform

In the last decade, we have observed an exponential growth in mobile devices which encourage the development of a variety of context-aware application. At the same time, it becomes a challenging task to integrate contextual information into application due to complex context acquisition and context processing mechanism. This increases the responsibility of context-aware application developer. The variety of sensors provides heterogeneous information. Temperatures, humidity, thermal energy, the water level in the river and so on are the examples of context information. Such contextual information can be used in various collaborative domains like health care, transportation, supply chain management, etc. The effective utilization of contextual information requires efficient data acquisition, processing, normalizing and reasoning techniques. So the application development platform must fulfil this requirement.

Celikkan et al. [2] proposed a novel, service-oriented context-aware platform. This approach is derived from observing the working of the operating system. The primary goal of this software platform is to provide relief by doing data management process easy so that context-aware application developer must focus on only application development rather than data management. In an open, dynamic and robust environment, it is required to design new methodology that helps in data acquisition, data management with efficient data processing mechanism which will be changed from situation to situation. This proposed framework separates context management and code of the application which handles contextual information on the behalf of the application. The PCAD platform is agile, robust and adopting changes without changing fundamental architecture. The PCAD architecture follows layered architecture. It is having an inbuilt security feature. This architecture can be used to define the industrial application. The focus of the development of these applications addresses the following issues:

1. Extensibility
2. Security and privacy
3. Simplicity
4. Generic
5. Service based.

This platform can be viewed as layered architecture in which platform acts as kernel, and applications are like user processes. The sensors are connected to middleware, and device driver is providing information to the applications. The sensor application subscribes themselves to the platform, and as soon as contextual information becomes available to the application, the notification is forwarded to the subscribed application. The application and platform are bounded by RESTful API which is lightweight and widely used in industry. If the devices are able to find the associated device driver, then we can work with a device which is similar to the operating system. This platform is scalable in which we can add or remove the sensor devices without any changes in the underlying platform. The platform enriched with a public interface which helps rapid development of the context-aware application. As we discussed above, PCADs follow service-oriented architecture in which subscriber can access PCAD's platform with its available services like Rule Service (RS). Data management service (DMS), alarm and notification service (ANS), reporting service (RS), security and privacy service (SPS), etc.

If we look at PCAD architecture, the PCAD platform acts as a bridge between sensor binding and application binding. Thread pool is another architectural component which confirms contextual information which is received from the sensor, this information is stored in proper format, and these applications provide required notifications in a timely manner. In pool thread, the task is assigned to the thread. Conceptually, each thread is further handles 32 subthreads. These subthreads are responsible for real-time delivery of raw data. The task of storage of data and its processing is also taken care of by these subthreads. This platform acts as a channel between sensor devices and application development environment. Livestock monitoring and smart parking are ongoing projects on this project [2].

The popular platforms for context-aware application development are:

(1) MoTwin
(2) Movea data fusion solution

These platforms are discussed in detail below.

1. **MoTwin**

The context awareness becomes a primary requirement of the mobile application as most of the users are connected to the Internet with the help of mobile devices. MoTwin is one of the leading context-aware platforms as a service provider in the market. MoTwin takes care of contextual information; this will act as middleware which resides between the application server and users' mobile device. It also takes care of all heavy operation and content delivery as well. With this platform, we can easily develop an application by integrating contextual information. It also provides the communication channel between the service provider and the customer. The best part of this platform is that it will help to develop a context-rich application by providing personalized information through messages contest and surveys. This will help to predict user's preferences location and actions. This platform helps to develop and maintain data when the user is not connected to the Internet. It is very safe, secure and privacy aware. This platform also supports the integration of backend services. This platform is specially designed for mobile application development. In this, data is partially stored on mobile devices and server. As far as the architecture of MoTwin is considered, it is distributed in nature. This platform helps to improve the efficiency and performance of the application [3].

2. **Movea**

When we are talking about context awareness, data acquisition and processing of same play vital role for application development. Movea is one of the top data fusion and motion processing solutions. This platform collects real-time information with the help of sensor devices and provides that sensor information to make a context-rich application. Movea is a developed novel, lightweight data acquisition framework and provides navigation for indoor scenarios. From this, we can interpret the accuracy of data. It also provides an embedded solution for a variety of mobile devices by considering the accuracy of data with low power consumption. Movea has designed its own data acquisition technique which will give higher accuracy in a real-time scenario. Movea has developed SmartFusion tool which makes an easy way of application deployment and application development for developers. The various applications are developed in a variety of domains like sport and fitness, smart home, mobile application development and so on. We can install SmartFusion studio on a personal computer or laptop. After installation, the mobile devices will be connected to Wi-Fi, and after connection establishment, all the sensor information of the mobile device are forwarded to the tool. This sensor data is stored in the file, and after the analysis of data, the context will be integrated by the application developer to make a smart indoor navigation system. The heterogeneity of data will be a major challenge faced by the Movea. Navy is planning to design a variety of applications including indoor and outdoor scenarios [4].

5.2 Context-Aware Applications: Design and Development

Designing and building context-aware applications require a completely different strategy as it involves a distributed collection of context from various places. So far, there are no conceptual models and standard methods to design and develop context-aware applications. In context-aware applications, the main challenge is to handle the heterogeneous context and to address the distributed nature. There is also the problem of distribution and reuse in the view of the design process. In fact, there is no standard software development process and model in software engineering perspective to develop applications using context.

5.2.1 Context-Aware Algorithms

Designing efficient context-aware algorithm in a distributed environment is very crucial and requires an understanding of the problem to be solved. In distributed computing use cases, the context is acquired from multiple computing nodes which are connected through local or wide area networks. Computing nodes are placed at multiple locations, and the computations are carried out at these multiple distributed computing nodes. Scalability, reliability, interprocess communication, resource management and fault tolerance are important challenges while designing context-aware algorithms. Specific to context-aware systems, the design of algorithms requires addressing context separation, interpretation, context storage and context acquisition with zero per cent downtime. For separation of the context, there are no standard algorithms, and this is the main reason why context is not integrated with many applications. But due to advancement in the computing and algorithm, today most of the applications are context aware. Many efforts have been made to handle context in the same manner as we handle user input. There is no standard querying mechanism to receive context from distributed and heterogeneous computing nodes. Context retrieval and its interpretation are not yet standardized, and in the sequel, designing context-aware algorithms is a big challenge. The routine desktop application receives input from input devices like keyboard and mouse. However, in context-aware applications, the context as input comes from the multiple remote distributed computing tiny devices like sensors and infrared emitters. Therefore, the requirement is that the communication for the context exchange should be transparent to entities providing context (i.e. sensors) and applications using this context. In such scenarios, synchronization in terms of the time clock, speeds, etc., are important parameters for the assumption. There are many context-aware applications which are critical like emergency medical services (EMS) and military applications where constant availability of context is mandatory. In such cases, a designer of the algorithm should be able to select appropriate data structures, suitable algorithmic strategies and design methodology. In addition to the issues presented and discussed above, resource and service discovery is also equally important, and as in context-aware applications, distributed sensors are communicating with various applications. The resource discovery mechanism should provide the required components for context access and the type of context information required.

Pervasive computing has the ability to bring revolution in the current technological environment. All the businesses are attracted to pervasive applications due to its applicability in a variety of domains. This technological area also faces challenges which become major hurdles in its implementation. The researchers from academia, industry, etc., coming together to overcome these challenges by proposing various solution approaches in the form of an algorithm or methodology.

When we are making transition towards implementation of the pervasive application, the time sensitivity has its own significance. At the same time, time-scheduling must be matched with the real-time scenario. Sometimes, it may possible that the contextual information can be used by the high-level application, and keeping another emergency contextual information in waiting for a buffer is not suitable for real-time systems. So to solve such problems in a real-time scenario, Fu et al. [5] proposed a context priority scheduling algorithm. This algorithm is based on the correlation between a variety of elements, priority and average priority of various semantic where the contexts is calculated which can handle real-time scenario efficiently. This scheduling algorithm reduces the average waiting time and improves CPU utilization. The pervasive application is a seamless integration of a variety of technologies and networks. In the last decade, wireless communication experienced a new era of connectivity paradigm. The wide variety of communication technologies improves the availability of resources over the Internet. The heterogeneity is one of the key characteristics of pervasive devices. These heterogeneous devices are roaming around the globe by connecting various access points without losing session information and data which is a requirement of the context-aware application. The user application should be able to switch between various active interfaces, which satisfy application requirements. Ahmad et al. [6] proposed an algorithm to fulfil the mentioned requirement with additional features. This algorithm makes the vertical handover decision process simplistic. This algorithm is suitable for the multimode mobile device. This proposed algorithm considers terminal- and network-related contextual information which makes an efficient handover decision-making process in the heterogeneous and ad hoc networks. The proposed algorithms are also scalable in nature.

The pervasive application always tries to understand the behaviour of the end-user so that the service provider can provide the best suitable service to its end-user. There are various parameters that showcase the behaviour of a group of the user who is having the same attribute or similar characteristics. Now, the recommendation engine becomes the most suitable tool to analyse the behaviour of its associated end-user. There are various recommendation engines which work on available static information in the small application. These recommendation engines are developed by considering commercial and open-source aspect. Zong et al. [7] proposed an open-source recommendation engine named as CARSkit. This is the first open-source recommendation library. This library is specially designed for context-aware systems which are implemented in the Java programming language. It is using an efficient data structure having a flexible configuration. There are various algorithms which are specially designed for the context-aware system. The security and privacy, mobility, addressing, personalization, resource allocation, etc., are the research areas where researchers are contributing continuously. In the near future, we can see a full-fledged context-aware application.

5.2.2 Design Process

To design any context-aware system, we need to follow the following phases.

Problem Specification
The first and most important phase of context-aware application development is the identification of the problem and its detailing. In this phase, we need to identify how contextual information provides a solution to the problem and how to use contextual information. When we develop any context-aware application, we need to make provision for context delivery to the user. At the same time, we need to focus on the presentation of information and the integration of a variety of services into the application. Once we deploy these services over cloud, we can make the automatic execution of services in the cyber-physical environment. In a real-time scenario, the user is the owner of various resources that could be related to storage, computing or networking. The tagging of contextual information and resource must be required to improve data retrieval or resource access time.

Data Acquisition
The second phase is data acquisition. In this phase, we need to select and identify the appropriate list of sensors which are going to be part of the system. Before developing a platform for data acquisition, we need to take a decision about how many sensor devices we are integrating into our application and what quality of data they are going to provide because the quality of sensors and sensor data will reduce the number of efforts required for data acquisition. Are these sensors independent embedded or

networked? What exact sensing parameters are required? What will be the exact location of the sensor device? What is a lifetime of the sensor? How often do we require data because sensors are continuously sensing environmental parameters? So while developing such application, we need to find out answers to these questions. Consider a scenario, in which temperature sensors provide the temperature at each second which is not required. There is no point to store each sensed value. It will unnecessarily consume the storage, in that case, we can store the value after every 20 s. Based on the application requirement, we need to decide how often we are going to generate associated reports.

Context Representation and Delivery

In real-time application, a variety of sensors are providing various values in a different format, so we need to create an application-specific suitable uniform format which can be easily accessible. Before storing data into a database or file, we must ask some questions. If we look at sensor data, it is an inefficient way to store data in the database, but in some cases, we need to use the database. Depending on our application requirement, we need to decide what database we are using. How we are going map sensor to a database like one to one or many to one. The time synchronization plays a very vital role because heterogeneous data comes into the database with a different velocity. The context delivery distinguishes the access mechanism and user preferences and contains another aspect to provide appropriate contextual information.

Reception and Action

This is the last phase of the context-aware application design. In this phase, the application locates the sensor nodes, which are relevant to a particular context. It could be in terms of service registration or service discovery. The user gets contextual information via queries, polls and notification. The frequency of retrieval of contextual information depends on the application. When an application makes a decision, it needs to perform various activities like data aggregation, calculation correlation, filtering of data and data fusion, and based on a situational judgement, conclusion will be taken out. Sometimes an application needs to use previously stored contextual information for efficient decision-making. Once we get contextual information, then the application can take a decision and perform an action. In this entire process, data preprocessing and heterogeneous data collection processing are very challenging. In a real-time scenario, when we develop an application and if the user gets wrong sensor value, then it is very difficult to find out the source of error because there are various possibilities due to which error can occur in the system.

5.2.3 System Issues

When we look at any context-aware pervasive application, it seems to be very easy; but there are a lot of issues that need to be taken into account. The pervasive application collects the contextual information about the physical world, it is a really challenging task to integrate such physical scenarios in the programming model, and there is a need to design a new programming model for pervasive computing. There are various sensors which are manufactured by a variety of vendors, and these sensors are having a different operating environment. Some sensors' data is unreliable which deviates the entire prediction mechanism towards wrong decisions. Due to the dynamic nature of the pervasive application, the traditional context-aware algorithms are not suitable for the pervasive application. At the same time, we do not have any recognized pervasive algorithm for application development.

Interoperability: Pervasive systems consist of a variety of sensors, devices and services. It is a challenging task for the application developer to manage all these things simultaneously. There is a need for a platform that supports interoperability of device, services, protocols, etc., if every service provider develops his own development platform without supporting interoperability system that becomes useless. Various devices are sensors providing data according to their convenience but become a critical task for the developer to handle the data due to the heterogeneity of data format. For rapid development of the application, we need a standard data format, protocols and frameworks.

Another important issue is related to middleware. The middleware carries various responsibilities like a collection of row data, end device management; translate entire data to user-specific format and investing its resources for proper execution of the application. Middleware needs to make some local decision in case of emergency or when unable to connect to the cloud server due to connection breakdown. So, it might be a good solution to decouple applications and context sensing. So that middleware can focus on a limited set of functionalities. While designing, we must take a decision about which computing paradigm is suitable for our application. The appropriate choice computing paradigm helps to reduce efforts. It could be either centralized or decentralized.

5.3 Context Awareness and Machine Learning

In context-aware applications and adaptation, a machine learning algorithm plays a vital role. As mentioned earlier in Chapter 4 of this book, machine learning is potentially applied to draw inferences about context information and to design respective decisions in the context-aware applications. This section presents a road map that enables all the stakeholders to exemplify the design and development of context-aware applications.

5.3.1 Knowledge Augmentation

The objective of machine learning is to gain knowledge out of the available training set and observation and integrate this contextual information into the system. Machine learning enhances the ability of the machine for decision-making. So it requires exact correct knowledge and applies that correct time. The real-time applications fetch different problem so it is required that extracted knowledge must build augmented and used efficiently for better decision-making. It is an iterative process, and learning is the most important aspect of knowledge lifecycle. Due to this, the capability of the system improves as the system faces similar kind of scenarios and takes decision smartly. When we collect data from the sensor node, we need to preprocess collected data. To design and develop a context-rich application, we need to collect contextual information extract feature out of that data which is suitable for a particular scenario and store that information with associated results. This stored result may help to take a decision when we face a similar kind of scenario. When learning started focusing on the semantics of knowledge rather than syntax, then the knowledge acquisition process becomes successful. In the real-time scenario, knowledge can be extracted from various sources like images, text documents, training set, videos and sometimes observations also. For knowledge acquisition, we can use a variety of machine learning approaches like a probabilistic, decision tree, etc., and knowledge augmentation prioritizes the data based on available metadata and puts value-added information at the top.

As far as knowledge augmentation is a concern, it is divided into four phases. The first phase is extracting interesting features of interest. In this phase, we collect data from various sensors, and we preprocess that information by removing noise and dirtiness of data. The quality of decision depends on the quality of the data. Once we get quality data, we start digging into data for the purpose of extraction of the interesting feature. In the second stage, the extracted features need to integrate into the application to make a context-rich application. In the third phase, we need to capture all the observations which are based on the second phase. The addition of current data in the previous data set helps to improve the quality of result with minimum time, and this process is called incremental knowledge building. Figure 5.2 shows the knowledge augmentation process.

As we are moving towards ubiquitous application development, the contextual information plays a vital role that there is need of seamless integration of devices, agents, networks and various services with contextual information which help to develop the context-aware application [8].

5.3.2 Machine Learning Approach

The proceeding and computation context-dependent information is an important part of a wide variety of application domains where the behaviour of the application is

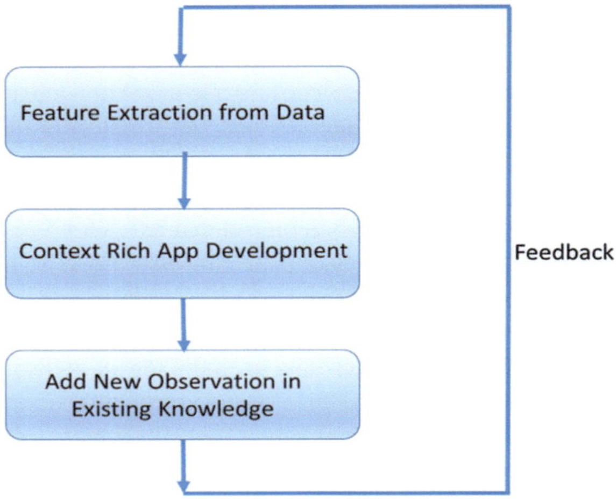

Fig. 5.2 Knowledge augmentation lifecycle

based on an execution scenario which is named as context-oriented programming. In this approach, the contextual information is provided in order to manage the scenario. The context-aware composition helps to improve the performance of the system by considering the dynamic aspect of the particular scenario. It also dynamically decides algorithm, and data structure improves efficiency. The learning is the ability to solve optimization and decision problem quickly based on the available data set. If training data set is not available in that case, we need to consider the dynamic observation of the system. There are many classification strategies and algorithms available. In machine learning problems, we cannot use the same set of solution for every problem. We need to modify the algorithm, strategy or policy each time as every strategy, method or algorithm have its own complexity, prediction accuracy, memory and computational overhead. In the resource constraint environment, we cannot use a heavy computational method or algorithm for resource constraint devices. The quality of the training data set determines the quality of decisions and prediction. The various approaches which are required to develop context-rich application are

1. Tree-based approach
2. Probabilistic

Various machine learning approaches for context-aware composition are presented and discussed below:

1. **Tree-based approach**

A decision tree is used for classification of data which makes decision-making and decision process easy. A decision tree is based on the inductive algorithm in which classification of each instance will be based on rules, and these rules are dependent on the application. In the top-down approach, each instance will compare to all internal nodes of the decision tree, observe each attribute of each node of the decision tree and infer a conclusion with the help of a decision tree. Consider a decision tree as a

Boolean function in which we are taking input as attributes associated with instance scenario and corresponding outputs 'yes' or 'no' decision value. In a decision tree, each node value is associated with the classification rule check, and branch represents the associated decision value. The journey of this decision tree which is based on learning started from ID3 reached to C 4.5 algorithm which is iterative in nature. The decision trees are based on the training subset. If the decision tree is unable to produce correct decision tree, then we add another training set, and this process will be carried out until we get the correct classification tree. In a real-time scenario when we want to compare unknown scenario, we start comparison operation from the root node. After that, we gradually move towards the leaf node by considering attribute comparison criteria. This process will be followed until we place an object in the associated class. A decision tree is the most suitable classification method required for the real-time application. There are various decision tree methods like ID 3, PUBLIC, C 4.5, CN2, etc.

2. **Probabilistic Approach**

It is a simple approach based on probability theory. Most widely used class of this approach is a naïve Bayes classifier which is developed on Bayesian statistics. Probabilistic classifier helps to predict the appropriate decision by considering the probability distribution function over the set of available classes. This approach gives justice for classification that can be useful in its own right. In naïve Bayes classification, the data set is divided into two parts, i.e. feature matrix and response vector. The feature matrix consists of the values or attributes of dependent feature, e.g. temperature, humidity, wind direction, and response vector consists of the output of associated scenario or we can call it as a class variable. For example, approval of credit card based on income and age output will be Yes or No. In this example, income and age represent a feature matrix, and output represents the response vector. The fundamental assumption of the naïve Bayes approach is that each attribute is independent and other is each attribute has equal significance. For example, in the above example, salary and age are independent attributes, and each of them having equal significance in the response vector. In Gaussian naïve Bayes classifier, there are continuous values and feature which is distributed by using Gaussian distribution. If we draw a graph of normal distribution or Gaussian distribution, it is bell-shaped and symmetric about the mean.

5.3.3 Case Studies

Context management is an important aspect of communication. When we read messages or email, we draw a conclusion out of it. Context helps us to determine something is right or wrong while developing any smart application. Context is very important. The user always expects the device should able to understand his/her context. In day-to-day life, we use various context-aware applications including satellite navigation. From the developer's perspective, we need to understand how our contextual information becomes a resource to improve whatever functionality we are going to deliver to his/her client. This context-rich application always helps to promote the best user experience. As the computing power of devices increases tremendously, it is obvious that the user is using a device with high computation power; communication capability expects the context-rich applications. In the current era, we not only adopt, but also store the same to improve decision accuracy. We are also trying to predict the next possible scenario of a user.

Adobe Experience Manager
The Adobe experience manager can use sling web framework which is based on REST principal. This web framework helps to develop a context-oriented application. In this framework, content is used as a context. The available bundles of sling help to develop an application with a different configuration. There are four implementations of the sling which is having its own significance in terms of context-aware computing. These implementations define the context, find the contextual information, store the context for future use and inherit some properties or attribute to another application. This implementation can be used by using the available API. These APIs are available in various programming languages including JAVA which provides the facility of resource lookup. This framework can be used for the development of Adobe experience manager. This tool acts as a bridge that provides personalized information of the client without considering the physical dimension of the device. It provides more security and scalability [9]. To make an application context-rich, this tool also includes artificial intelligence and machine learning approach. It helps to uncover the user's preferences and bring the best information and services to them. The context intelligence can be provided by giving the right information at the right time. Use case like Adobe experience manager is elaborated in the last part to understand the operational functioning of context-aware applications.

5.4 Conclusions

Context-aware computing and development of context-aware application are not yet fully matured. It needs further research and standardization to lay down foundations. First, we have defined various approaches and requirements to develop context-aware systems which include scalability, security, abstraction and resource constraints. In the sequel, the standard reference model with various layers is presented and discussed. Due to lack of standard conceptual framework and methodology, the next part of this chapter focused on the design and development of context-aware applications. The requirements to design context-aware algorithms, design process and various system issues are also discussed in detail. As we know, machine learning plays a vital role in building context-aware applications and productâ€"the last part dealt with how machine learning algorithms are crucial to design applications using context.

References

1. Chen, H.: An intelligent broker architecture for pervasive context-aware systems (2004)
2. Celikkan, U., Kurtel, K.: A platform for context-aware application development: Pcad. In: 2015 Federated Conference on Computer Science and Information Systems (FedCSIS), pp. 1481–1488. IEEE (2015)
3. www.motwin.com, https://www.businesswire.com/news/home/20120918006446/en/moTwin-Releases-New-Version-Context-Aware-Mobile-Application
4. https://www.crunchbase.com/organization/movea
5. Fu, X., Lv, Z., Huang, W.: Research on context-aware scheduling algorithms in pervasive computing environments. In: 2010 5th International Conference on Pervasive Computing and Applications (ICPCA), pp. 277–282. IEEE (2010)
6. Ahmed, T., Kyamakya, K., Ludwig, M.: Design and implementation of a context-aware decision algorithm for heterogeneous networks. In: Proceedings of the 2006 ACM Symposium on Applied Computing, pp. 1134–1138. ACM (2006)
7. Zheng, Y., Mobasher, B., Burke, R.: Carskit: A java-based context-aware recommendation engine. In: 2015 IEEE International Conference on Data Mining Workshop (ICDMW), pp. 1668–1671. IEEE (2015)
8. Kulkarni, P.: Reinforcement and systemic machine learning for decision making, vol. 1. Wiley (2012)
9. https://www.adobe.com/in/marketing-cloud/experience-manager.html

Chapter 6
Context-Aware Computing: Telecom Perspective

Abstract A pervasive computing system has been used in several domains. This chapter elaborates on its use in the telecom industry. The mobile user needs various information like traffic updates and weather conditions. From the perception of mobile users, this chapter begins with the layered architecture of context-aware system. The interaction of the telecom industry with the media context has been increased in the last few years. This chapter discusses several context-aware services offered by the telecom industry. The services include content, location, session and identity. The chapter continues to discuss telecom SOA architecture. The chapter finally concludes with a discussion on a context-aware business model which includes perception from the user, network and third party.

6.1 Telecommunication Context-Aware Services

Context-aware systems and applications are the modules which are specifically used to collect information/facts from the surroundings at some point in time. The context-aware computing system is used in hardware and software to collect and analyse the data automatically. The data/information could be related to the electronic device, a person or any application which includes the location of a particular device, user, time, nearby device or person, activity, task, identity, role, etc. Many data sources are devices of context-aware computing or application, like sensors, cam-cords, world-wide positioning satellite receivers and web browsers. Context-aware computing systems collect information or data from these devices and other sources and respond according to pre-established rules or computational intelligence [1]. Context-aware systems for user applications can guide many services as well as enhance user experiences by incorporating features such as reality in visual and display, contextual marketing messages and context information delivery. A context computing system is defined as a property/ability of an electronic device (e.g. Mobile) to present the proper information to the other associated user. It operated as a technology driver for machine-to-machine (M2M) and the Internet of things (IoT), event-driven context computing environments.

P. N. Mahalle and P. S. Dhotre, *Context-Aware Pervasive Systems and Applications*, Intelligent Systems Reference Library 169,
https://doi.org/10.1007/978-981-32-9952-8_6

Telecommunication context-aware services play a significant role in computing technology, which collects information about the present place of a mobile user to give more applicable services to the user, e.g. real-time traffic updates and weather updates. This information can be updated with the help of communication devices, users, mobile sensors and applications [2]. Nowadays, telecommunication context-aware services are used to incorporate one of the core contextual capabilities exposed by the telecom service provider as part of the business service [3]. From the functional point of view, a context-aware system is represented as a layered perspective, the lowest layer being the sensor layer, followed by raw data retrieval layer, pre-processing layer, storage or management layer and finally the application layer. In this layered structure, the context management system (CMS) is fully responsible for retrieving raw data from sensors, abstracting and combining the sensor data into the upper-level context. The data is made available to the context-aware applications only after the refinement done through the underlying layers mentioned above.

The layered framework for context-aware systems is depicted in Fig. 6.1. The bottom-most layer, i.e. 'sensor layer' is responsible for retrieving data from the environment (e.g. user device, used access network or social network). The second layer is the 'raw data retrieval' layer extract data from the lower layer (sensor layer) using specific APIs or protocols. The third layer is the preprocessing layer; its responsibility is to reasoning and interpreting contextual information received from the raw data layer. It transforms the sequential information from the fundamental layer to a higher abstraction level. The fourth layer is a storage space and organization layer which organizes the gathered data and makes them available to a third-party's application in a synchronous or asynchronous way.

The fifth layer is the application layer, wherein the responses to context changes are reflected, for example, displaying text in enhanced colour and contrast if enlightenment turns bad. The purpose of a context management system (CMS) is to pro-

Fig. 6.1 Layered framework for context-aware systems

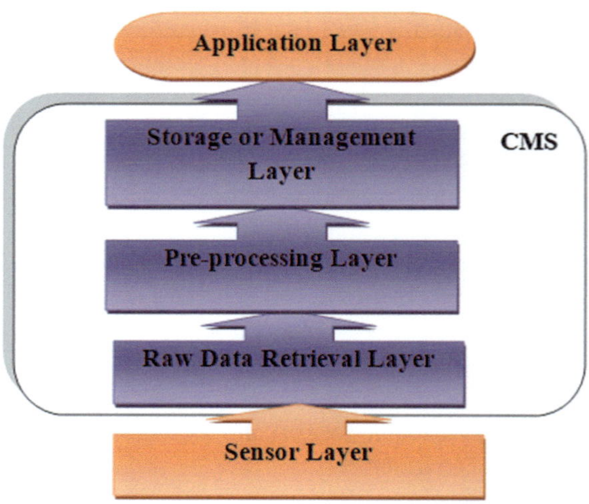

vide the capability for various users with different permission levels and to control telecommunication context-aware services/content.

Key features of CMS

For quite some time, the IT business has been excited by the possibilities of both portable mobile devices and pervasive processing. Gadgets like PCs, notebooks, PDAs and advanced cells have just liberated us from the technological shackles of workstations and work areas. Vast scale experimental computing, such as the cloud and grid technologies have given support of an exceptional number of PC clients [4]. A fascinating aspect of mobile computing is the decoupling of functions from the area where it is performed, as shown in Fig. 6.2. Users can maintain accounts or records, get business information, exchange emails and interface with colleagues continuously from home, at the air terminal, in the workplace, on the shop floor and many more [4].

Major Applications of Context-Aware Services

There are many real-time applications of context-aware services that will affect our daily life in all its proportions (e.g.—at home, at work, in public spaces, etc.). It has certainly provided a way for users to familiarize with the working of applications in order to achieve user expectations and it also facilitates the way by that user can specify actions needed to be performed to under particular situations, based on the context. This facility of service adjustment can be useful and may improve different tasks such as selection of services, adaption of privacy and security parameters based on the situation to achieve strong access control mechanism for particular scenario, selection of communication protocol to optimize communication overhead and semantic analysis of user request to retrieve parameters which can improve result of the request, i.e. content adaptation.

A few typically predicted context-Aware applications

Following are a few commonly predicted context-aware applications which include location-based services (LBS), context-aware communication (CAC), context-aware buildings and context-aware recommendation systems (CARS).

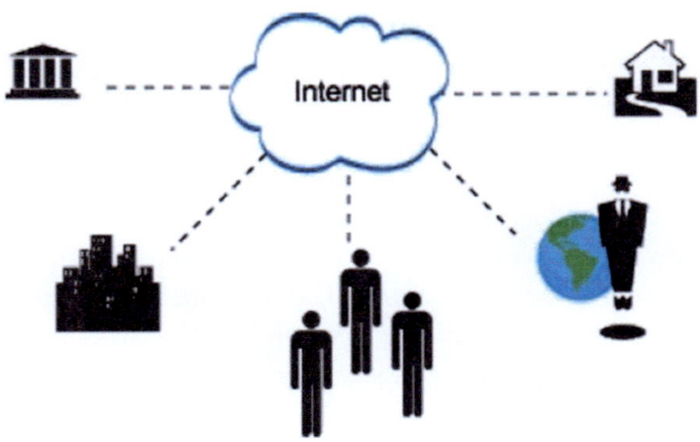

Fig. 6.2 Mobile computing decouples function from location

- **Location-Based Services (LBS)**: in this type of services, location is the prime context for service selection. In location-based systems, services are categorized on the basis of the location.
- **Context-aware communication (CAC)**: These applications work by applying the user context to minimize communication barriers. Some examples of context-aware communication applications are nonverbal and electronic communication services (SMS, MMS, chats, email, electronic message boards and mailing-list), these are presented below.
- Service that can check whether the urgent message is delivered to the recipient, it can also check whether it has been read by the recipient.
- Leaving messages at certain places (not to a specific person) making it available to anyone who arrives at the same place, similar to an electronic press release board;
- Service that can notify the user regarding received messages based on the user's context, e.g. here, context can be timing or location. E.g. based on user's timings of coffee break such notifications about the reception of the messages can be done when the user is in coffee breaks.

6.1.1 Content Services

In the last ten years, we have seen that there is an increase in the interaction between the telecom industry and the media content providers, each of them trying to establish their individual markets. With the ever-increasing number of customers worldwide, the demand for media content has also increased, which requires additional network bandwidth to augment the quality of their media expenditure and interactive experiences. For the telecom engineering industry, this meeting with media content providers is all about the carrier moving beyond the pipe to where the new money is a business surrounded by media content making and content release are naturally tied [5]. Real-life content service provider system is depicted in Fig. 6.3.

With the consumers having their own choices of smartphones and wireless plans, the future growth for the telecom industry lies in enabling them to view TV channels, movies and other media via all those mobile devices and channels. This convergence with media content providers allows telecom carriers to offer connectivity and premium content as a whole package. Satisfying customers do not only depend upon the contents delivered but also depends upon other features such as timeliness and accuracy in delivery of content [6].

Concerning communication services, commercial enterprise users may also, every now and then be overloaded with work making them temporally unable to deal with incoming communications. The fundamental context sources are user's calendar and usage history of various services. Mobile carriage carrier vendors have commenced supplying top-class rate services through a devoted sequence of numbers, the use of the short message service (SMS) and the multimedia message service (MMS). The service provider provides value-added facts and enjoyment through SMS and MMS

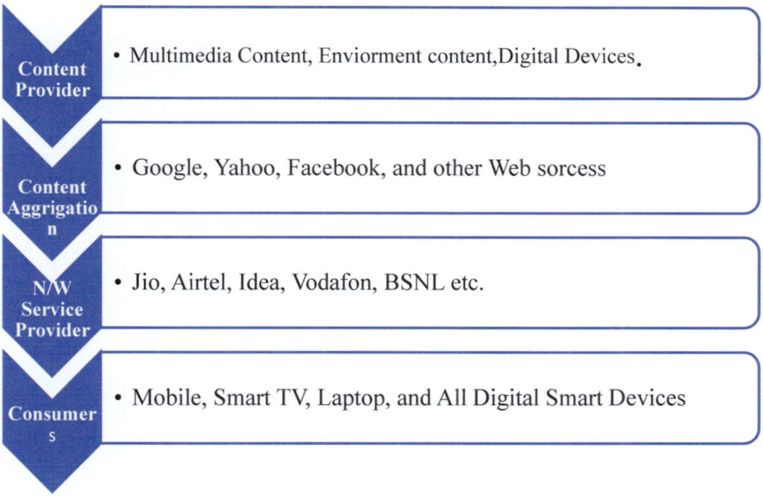

Fig. 6.3 Real-life content service provider system

to cell devices. SMS and MMS can be used for a large range of purposes, which include the following:

1. Interactive 'chat' services	2. Voting on television shows
3. Competitions	4. Ring tones
5. Weather forecasts	6. Astrology services
7. Banking services	8. Online shopping status report

SMS and MMS are also anticipated to promote the improvement of mobile commerce, enabling the purchase of less expensive commodities such as bank account information via SMS. Customers typically receive a bill from their mobile access service provider on behalf of a third-party content provider, wherein the content provider shares the returns from the content service.

Content Management Architecture

The architecture presented in Fig. 6.4 gives an idea about the content management strategy. It helps us understand the fundamental concepts of content being managed, which is actually available for all subscribers. It has an internal content repository, and it provides a storage area for real-time (live) contents. It also provides different functionalities to the two repository patterns for the identification of shared content.

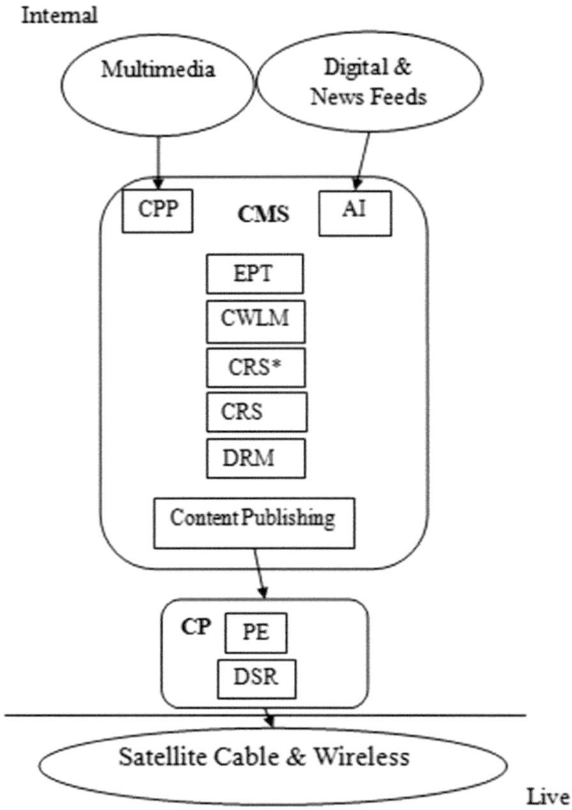

Fig. 6.4 Content management architecture

The live storage area is an important part of a retail outlet in which millions of consumers may access a limited set of available contents [6].

Abbreviations used in architecture

CPP—Content Provider Portal	**AI**—Automatic Ingest
EPT—Editing and Preview Tools	**CWLM**—Content Workflow and Lifecycle
CRS*—Content Repository Services	**CRS**—Content Rendering Services
DRM—Digital Rights Management	**CP**—Content Publishing
DSR—Device Specific Rendering	**PE**—Portal Environment

There are many key elements in content management systems which are explained below.

1. **Content Capture**

Content capture handles all content types like, text, images, audio, video, XML and code or HTML that are associated with metadata. Here, in this case, the content provider will allow third-party vendors to push contents anywhere [6].

2. **Metadata Management**

It is very useful for managing and configuration of available contents in two categories, i.e. active and inactive.

3. **Content Workflow and Lifecycle Management**

With a large amount of involvement in the process and the fact that the data change their nature while being processed through various phases, content workflow and lifecycle management take a prominent place in the content management architecture in order to identify the workflow and act accordingly.

4. **Content Distribution**

The purpose of contents distribution is to carefully handle the contents when it is physically in transit from one location to another.

6.1.2 Session Services

Session services help to provide seamless connectivity during a communication between a client application and a service. In communication, disconnections might happen due to the intervals in connectivity. To overcome disconnections, session services offer external or supplementary functionality of masking the disconnections. We may also say that session service is the communication between the client application and the service to finish the customer's task. In this, we focus that the communication is between the client and the service and it is not related to the particular service provider or process. Precisely we can say that customer is allowed to switch service providers if needed, provided that the new service provider offers the necessary functionality, i.e. the service providers can be changed but not the service [5]. We can achieve this in two ways.

In a first way, the client can connect to a new service provider if the current service provider moves out of communication range and can use functionalities of the new service provider. While switching, it is kept in consideration that new service provider is offering functionalities similar to the previous service provider, this switching can happen automatically when a client comes in communication range of new service provider. In the latter case, the client gets service by the host-to-host migration, like 'follow-me' session. In this approach, process that offers a service migrates from host to host. This follow-me session provides continuity of service and handles all challenges related to it. This process uses context-sensitive biding approach and location agnostic communication protocols to keep migration process transparent to client host and client application. The context-sensitive binding is a method of the strong process migration. User can user mobile services and applications for varying context information and surroundings. User may get contextual information when services are made context-aware.

Background And Context-Aware Session Services Management
Service-oriented computing has the feature of decoupling which makes it applicable to every domain. The interface of functionality and its implementations are not strongly coupled hence services can be advertised by service provider's run time and user can discover these services runtime. The proxy-based and web services are the two approaches to designing service-oriented computing architectures.

The working proxy-based service-oriented computing architecture includes the host machine on which the server process may run and the proxy-based object is required to work as a remoter handler. The proxy-based object functions as the server at the local level for client requests. The interaction between client and the proxy-based object is similar to general interaction with local objects. The functionality of the proxy-based object varies based on the requirements of the client, in some cases, the proxy-based object can fulfil client's request by its own, however, in some cases to it need to connect the server that is active on the provider's machine. While it connected to the server on provider's host machine, it gives the impression to the client that request is handled locally. The communication between the server and proxy is played a vital role in the implementation of the middleware as the implication of the mobility between proxy and server is handled by middleware. This mobility between both ends of communication needed to be handled for seamless interactions.

Web services use lower level languages such as Web Services Description Language (WSDL) to describe the way in which data is sent through wires. To know 'what data is sent through cables and why it is sent' higher level languages are used. The user request is matched with web service description if it is matched then user request can be answered, for this high-level language should be concise for easy matching of two entities. At lower layer, Extensible Markup Language (XML) is used to provide rules to encode the description documents. Nowadays with the increasing popularity of web services, matching is not only restricted to syntax, but it also goes beyond it and the semantic of the client request and web service description is matched for better results. For semantics analysis ontology can be used, the high-level ontology languages retrieve semantics from both entities. The DARPA Agent Markup Language (DAML) is used to construct ontology, and Ontology Inference Layer is used to interpret construct on and ontology is often by itself structured into layers. The Resource Description Framework (RDF) can be used for the construction of the ontology; however, it is not powerful to describe complete ontology. To describe complete ontology, DAML + OIL provides more constructs and relations.

6.1.3 Location Services

The physical context of the user or user's device is the most important factor which will help in collecting details about the location of a mobile user. The information that can be retrieved from the user's context is depicted in Fig. 6.5. Nowadays, mobile computing context is not restricted to the user's location and time, it can be other facets of the user's environment such as device characteristics of a device like

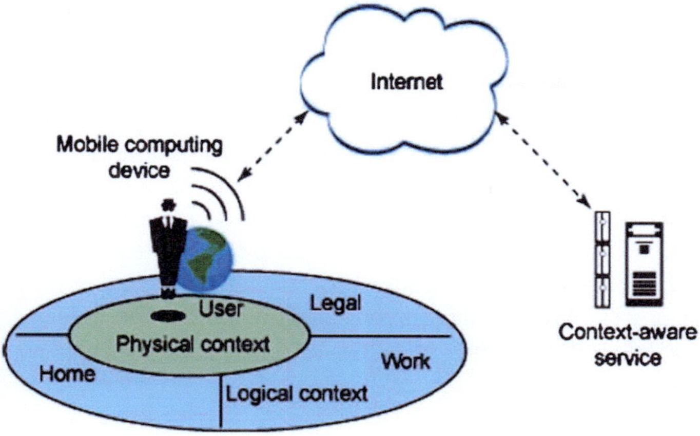

Fig. 6.5 Mobile user context

network identity, screen size, keyboard configuration, communication capabilities GPS sensors and many more [7].

In more common terms, we can refer context as any aspect of a situation where an entity (person, place, or computational object) may invoke computational functions. Today's digital era mobile systems and its functions are referred to as calls for service from either local or remote service providers. In such type of scenario, service consumers are the mobile user and his device [2].

6.1.4 *Identity Services*

Identity services help in realizing real-time authentication and authorization of service providers and users across heterogeneous domains (cross-platform domains), thereby building a trusted relationship between them [2].

The way of identification of context, based on current identity management methods, helps to deliver an unbelievable form of cleverness, which is applied to current network capabilities. Context-Aware Identity Management is a powerful way to help any service provider that influences the enormous assets of context-aware information. This also helps to make identification of users as well as devices (and allow them access to authorized services) easier and safer from vulnerable events or threats that can be harmful to users and devices.

Context-aware information refers to dynamic, situational and event-based data such as location, proximity, environment, access network type, user profile and many more. Service identity management refers to the way the service provider's identity can be identified and also monitor its authentication in regard to the wide range of users and devices based on their authorized use.

The identity service is one of the major services that can help in identifying provisioning and trust which is a sort of validation technique/process. Many identity providers use strong biometric security techniques for authentication that are more trusted than other available security mechanisms for authentication, which generally use only username and password for authentication [5]. For strong security measures, identity providers support more than one type of authentication, e.g. we can also relate it to identity provisioning for trust value; the trust value is specific to the communication sessions. The identity provisioning factor of trust may vary based on the parameters, e.g. the user registration process, i.e. Google and Yahoo allow user registration process easily without strong identity verification methods, however, bank, university, etc., generally not allow user registration without verifying the identity of the users [2].

6.2 Telecom SOA Infrastructure

The service-oriented architecture (SOA) is an architectural representational style. To use existing available software applications and systems, there is a need to have loose coupling among different software applications that are interacting. SOA provides such loose coupling between interacting applications. This feature facilitates the implementation of modern business services that can be easily reused and integrated with other services to complete the given task with assigned dedication, with respect to the failure and increasing adaptability of it, thus creating an elastic and amendable infrastructure [2]. Using the SOA architecture within the reach of an organization enables it to utilize more resources on innovation and on delivering new business services. A system that can successfully integrate SOA infrastructure can minimize the disturbance of planned or unplanned outages in an enterprise.

The telecommunications industry has witnessed many changes in the last few years, leading to advancements in its features, resulting in rapid growth. Telecom carriers over a long period in the past have created a multifaceted web of interconnected applications [8]. It is difficult and costly for newly introduced systems to interface with these legacy applications. The interfacing with legacy applications is the major issue of today's wireless communication world; this can resolve by the well-designed SOA which can address all problems related to the legacy applications. SOA connects the application to application seamlessly without a point-to-point interfaces; it facilitates systems to connect each other without boundaries [9]. The telecommunication industries can earn more money and recover revenue leakage by creating a reliable end-to-end communication.

Unique Aspects of SoA to Telecom Business

Subsystems in the telecom industry do not provide importance to the core business functionality but rather concentrates on enabling the interaction among the core applications of business, their partner systems, their components other available unit systems to deliver one or more integrated business systems. Accordingly, SOA is mainly focused on the details of communication between these endpoints [8]. An SOA solution defines a framework to the innumerable available extents, which helps in achieving dedicated enhancement. It achieves the fundamental performance by implementing the following mechanisms:

1. **Smooth the progress of Interaction between Endpoints**—the SOA aims to enable interaction between applications that are different than each other to complete the assigned task with respect to the interaction between two points [9].
2. **Enable Distributed Operation of Business Components**—for better usage of business components, SOA should have the facility to use them in a distributed manner through the network. This is very important for the responding situation so that it can be beneficial for both situations with heterogeneous components, and homogeneous components, i.e. components that nearly work well together as a set of dedicated links and connection between them with the help of SOA architecture give overall idea about any situation and handle it carefully in the form of heterogeneous or another way.
3. **Padding Endpoints**—In the ideal case, the communication endpoints need not know about a business object like information about consumers, their messages and the way in which their request can be answered. To achieve this, various business applications are implemented in such a way that their context can cover all business requirements; this way these applications can fit the new business world [9]. In the representation of the business features, a well-designed SOA infrastructure can play a very important role; it can present business functionality in a well-defined manner. That is, the location and provider of the service do not need to be known by the requester of the service. This information can be easily understood with the help of SOA architecture for the enhancement of business service requirements.
4. **Provide an Infrastructure that is Extensible and Scalable**—The well-designed SOA infrastructure should facilitate to add new endpoints and it should be flexible to handle increasing workload, thereby making it beneficial for requirements understanding and developing solutions to the requirements within the defined scope.
5. **SOA Enables the recycle of Endpoints**—Every application and its associated endpoint should have only one crossing point for all interactions, or in some cases, number of crossing point should be minimal. To provide simplified yet versatile SOA design, it should have conventional interface point to add new interactions. Every business system has some requirements from SOA. Security, application management and directory organization are the prime requirements. SOA should have some measures and internal access to infrastructure services to achieve these requirements.
6. **SOA Infrastructure Enables a Phased Approach to Implementation**—The SOA communication solution should enable easy implementation of the short-term solution in an incremental fashion and put down the foundation for the

accomplishment of a complete architecture. Service-oriented architecture (SOA) follows the architecture style, which aims to achieve a combination among interacting software application. This architectural style helps to benefits from the investments that are done on the existing applications and systems. This SOA style facilitates flexible and adaptable infrastructure by providing a way of implementation of modular business applications that can be seamlessly integrated with other applications and can be reused. Due to these features of SOA, style organizations can focus on budget and resources. It also helps the organization to deliver new business services by overcoming different planned and unplanned risks.

Advantages of SOA are
The advantages that come with using SOA infrastructure are listed below [9].

1. **Reduction in development time and cost**: SOA style facilitates easy integration with other services to deliver a new one and it can be easily reused for complex applications. This feature provides better enhancement of business applications.
2. **Lower maintenance price tag**: Reusable services decrease the amount of internal difficulty in projects of enterprise services.
3. **High-quality SOA services**: SOA provides greater than before service reuse which helps to create better services by applying many testing cycles. For better quality services, these testing cycles are composed of various service consumers.
4. **Lower incorporation expenditure**: To connect various applications seamlessly and quickly, services need to work together. This is very well known by the consistent services.
5. **Reduce the risk tag**: Risk can be bridged with the help of SOA transportation for communication. Reusable services provide greater control over corporate and IT supremacy policies, and decrease the overall compliance risk to an enterprise.

Business Scenarios—Telecom
SOA can be implemented by a phased approach which provides benefits to particular scenarios in the business areas, which are listed below. For example, fraud management. It describes details of service application implementation, i.e. its granularity, implementation model, governance and choice of partner. These can be applicable to the other business scenarios as well.

Fraud Management
The IT industries lose its revenue due to counterfeit attacks; this is a major issue of concern for SOA infrastructure. The IT infrastructure operators use the Fraud Management System (FMS) to minimize the revenue loss; it plays a vital role in the IT industries. Their approach is to recreate a role for both operator and subscriber. It recreates roles while new subscriber signup and subscriber signup for usage of any service provided by the operator. Service operators need to do huge investments to take support from FSM for communication between subscriber and operator. The

operator can use off the shelf FSM or they can customize as per their requirements, however, investments are required in large scale in both the cases. The FSM can be used in entire telecommunication industries with the accomplishment of SOA principles in the FMS. It can provide the capability to the operator to extend their Fraud Management services to other operators/MVNOs. Furthermore, it also helps to introduce changes in the business process and rules, across the infrastructure in a seamless manner [9]. Figure 6.6 shows a typical workflow in an FMS.

On the other hand, one of the other benefits of adopting SOA in the FMS infrastructure is to deliver the technical enabler for reusing the FMS across the IMS infrastructure.

The SOA Infrastructure Service Way

The SOA can be implemented by applying the evolutionary process, as the future services will be run parallelly which can lead to reuse of applications. This approach of revolutionary services does not mean to have huge efforts. This revolutionary approach means to define step-by-step process to achieve the desired aim. This process may be progressively superior, and it may address a wider range of scenarios, as each successful installation is completed and seen to work. SOA infrastructure services are presented in Fig. 6.7.

1. **Model Business Tasks and Processes**

FSM may use its own business process or it may adopt the process of other business to achieve its goal. In both cases, FMS has to revise the business process to satisfy business requirements of the particular IMS infrastructure. By this way, FSM services can be used by the third parties that are outside of the network of operator [9].

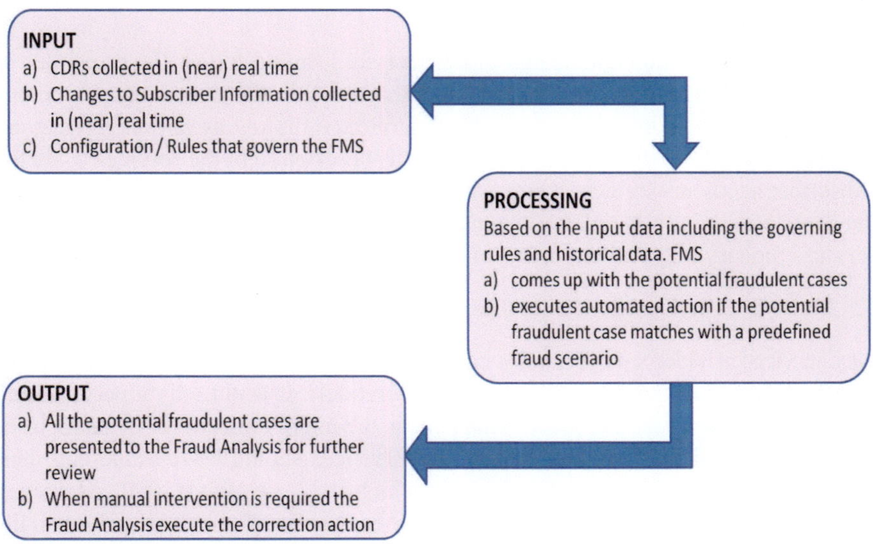

Fig. 6.6 Typical workflow FMS

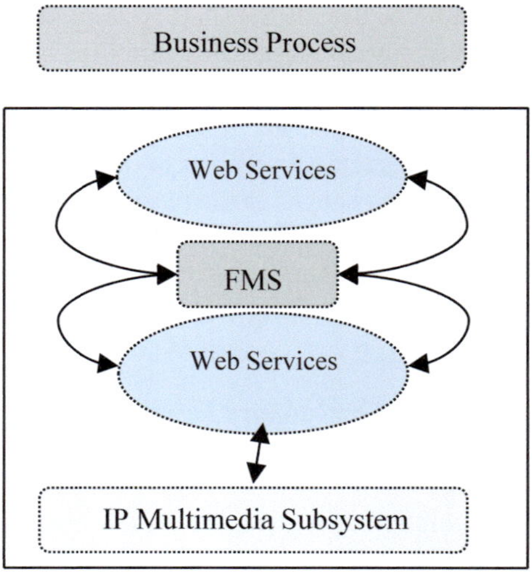

Fig. 6.7 SOA infrastructure services

2. **To Create Services for Functionalities offered by FMS**

To complete the particular business task, the SOA infrastructure can use services from existing functionalities. As mentioned above, the application functionality that can detect fraud and can correct is modelled as 'Fraud Detection and Correction' service. In the industry for some task, the functionality may not be implemented; hence, there is a need to define new service that can implement the functionality which is required. Concisely, we can say that we are in the integrated world, where the subscriber/user might be connected with many contributions from the operator. Hence, there is a need to create new services to hang up all contributions to provide desired functionalities [9].

The implementation and design of new services should follow the principles of SOA. It should be interoperable, loosely coupled, reusable, discoverable, autonomous and compostable. At this stage, a registry of services is the key, as various services are being developed [9]. The registry maintains the description of the services when the subscriber needs any services it requests for the same and this request is matched with the description of service which is maintained by registry hence for the discovery of services registry plays an important role [9].

3. **Make the Environment Scalable**

Any service provider's performance can be evaluated on the basis of how exact service is provided to the subscriber as per their request. It is not only a performance parameter but how fast this process happens is the major concern for evaluation for any provider. For the betterment of system, services should be reusable and their compositions should be dynamic for this which may use heavy XML. However, these lead to latency issues. Furthermore, there is a need to give more attention to the security issues as services are exposed to external parties. The system's throughput levels can be predicted based on its usage and transactions. This can be helpful to determine classes and types of products can be used hardware and software enterprise architectures. SOA facilitates to add hardware in the system as per increasing requirements, and it is the most important advantage of the SOA.

4. **Implement a Rich Internet Application based on the presentation layer for the Fraud Analysts**

There are many web applications that have the characteristics and functionality of long-established as well as customized desktop applications in the market. All these web applications have numerous dealers who provide solutions to provide RIA. It is not a mandatory requirement for SOA infrastructure. FSM have governing rules and these are configured dynamically and efficiently by the FSM using a thin client (web browser), in this RIA can assist to the FSM.

Governance Rule for SOA Infrastructure in IT

The purpose of SOA is to reduce the conventionality between applications in IT infrastructure and to deal with it, the government has given prime importance and more attention. There is a need to implement federated structure as miscellaneous users are now reusing the same assets with their varying priorities [9]. The development of repositories of enterprise helps to define services available and it requires various facilities, tools and products. Using these tools, services can be defined and described with standards by which it can be seamlessly discoverable. Additionally, the enforcement and definition of the policies are also referred to as governance. The execution of the business processes like the level of auditing, securities, is also referred to as governance, etc. It is important to form a governance body which will also include users in the organization not only include products of the organizations. Such governance body will be the one centre point for defining and monitoring the '*rules of the game*' so that they can easily allow or express the needs.

The SOA Infrastructure for Granularity of the Services

The granularity of the service is the important parameter in the design of SOA-based solutions. There are different requirements for designing SOA solutions based on the system, e.g. the Fraud Management System, required SOA implementation should be revolutionary, and it must provide benefits at each milestone. Concisely, we can say that in initial stage system services can be coarse-grained and with the time for it can be converted to the better-grained service for efficiency. This is one side, however, on the other side, at an early stage of analysis, there is no need to process coarse-grained services. Hence, there is a need to keep stability between the services with respect to the granularity for better performance of services.

The following important four major functionalities presented by an FMS are:

(1) Detection of the fraud.
(2) Investigation of the fraud.
(3) Prevention of the fraud.
(4) Measurement of the fraud.

Commonly, SOA implementation at early stages can be modelled approximately using functionalities of common-grained services. In succeeding stages, superior grained services can be formed by drilled down these functionalities. Let us take example 'Fraud Prevention' and let us examine how the services will be modelled in excess of time.

Figure 6.8 presents the primary functionalities offered by FMS. The business procedure has the elasticity to select service which will invoke exact functionalities that is necessary to complete the task by providing all necessary functionalities. This service can be coarse-grained at the initial stages. Further, by sorting out the 'Audit Trail' functionality into a split service, it enables its recycling, as it is required by not only 'Fraud Prevention' but also by 'Fraud Detection' and many other functionalities too [9]. As many services are broken down into their granular levels, its capability to 'mix and match' for rapid growth and its ability to reuse has been accomplished successfully.

Choice of Partners and Implementation

It is important to decide the appropriate partners from both ends, i.e. from a service and product. It is an evolutionary approach to select service by development partner's point of view in the SOA World. There are many products that address similar requirements at different levels, and it is important to address these needs. For example, authentication and authorization functionality may be already implemented in the web application server; these may be implemented at a different level than the required one. High-level view of SOA execution at partner office is presented in Fig. 6.9.

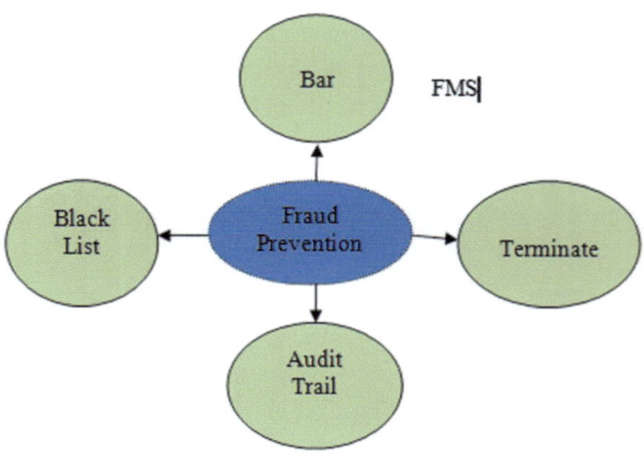

Fig. 6.8 Primary functionalities offered by an FMS

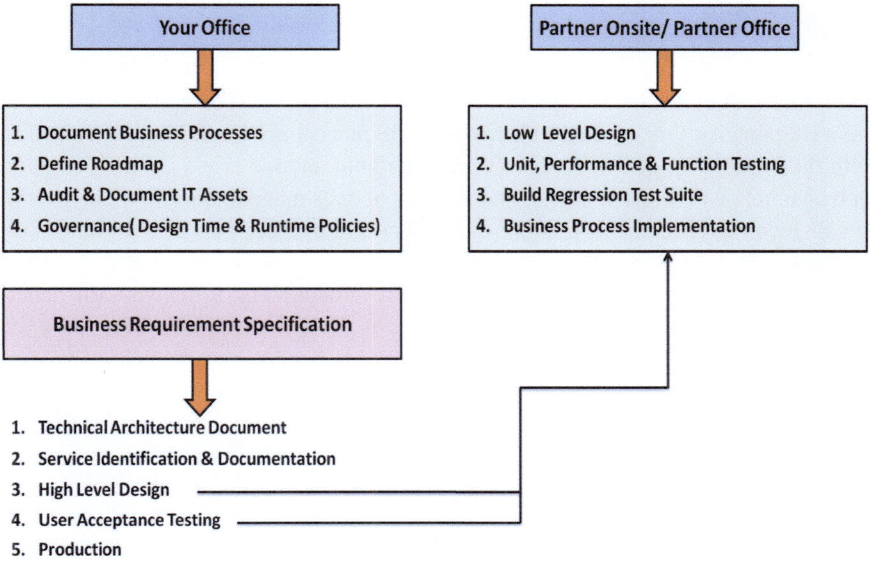

Fig. 6.9 SOA execution at partner onsite/office

Offshore service partners with their credentials on SOA infrastructure and the right size for agile and effective customer service would be good fits for the purpose of implementation. The purpose for offshore is to carry in both the cost-effectiveness and value propositions as enterprises which are looking for partners who will communicate their business goals and go the extra miles required. For an event in SOA execution, execution testing is done ahead of time in the advancement cycle to provide food for early discovery of inertness and bottleneck issues. In the accompanying graph, the segment set apart in the translucent spotted region speaks to the iterative procedure that will be included while conveying the administrations for a distinguished business process.

Key Benefits

Apart from bringing down working costs, this mechanism will give a one-stop answer for the difficulties presented by the expanded movement on the system and furthermore by the expanded and new abilities of the fraudsters. By architecting the arrangement in light of SOA, the coordination between the FMS and IMS can be made consistent. It helps to facilitate new business rules and security channels can be connected at the same time over the BSS and IMS layers successfully and effectively.

Customer Management

The core part of the BSS layer is Customer Management System (CMS). In CMS, cost to provide service and customers viewpoint are two important areas of focus. These two points can improve business; the customer viewpoint can help in increasing customer satisfaction that can lead to high sales and the cost to provide services help to lower operational cost.

1. **Cost to Serve**

The type '*Cost to serve*' mirrors the cost to serve and bolster a supporter. Bringing down this cost expands the capacity with utilitarian devotion to upgrade and increment its capacity. One of the roads to diminish the cost is by enabling the endorsers with the capacity to deal with their records without anyone else's input through the web channel or IVR and lessening the number of calls made to the CSRs, supporting in call focus advancement. In spite of the fact that a larger part of the administrators have self-benefit fused through the web channel, acquainting corrections or new administrations to understand the more prominent capability of the web self-benefit is still a test for time to market and high IT costs. This is the place where SOA will profit.

Example: Consider two business firms in client administration and perceive how they can be displayed in the SOA framework:

1. **Registration of new subscribers**
2. **Subscriber account management**.

As portrayed in Fig. 6.10, the two business processes share common services that are Update Profile and Update Billing Information to address the functionalities of a particular process. With advancements in accordance with the concurrent arrangement, the business forms summon both the remote and broadband administrations to give the client an entire arrangement without managing various enlistment cycles. Presently, if the Showcasing Group concocts an advancement, wherein all compensating month-to-month supporters who set up coordinate pay or auto-instalment are qualified for an extra heap of instant messages, at that point Refresh Record Subtle elements would need to simply conjure an extra previous administration Refresh Packs.

Therefore a common service across the channels (Web, CSR, etc.) to the reuse potential is very high. Security policies can be distinct to control access to the various services through these channels.

With the foundation benefits set up, the conceivable outcomes of making and altering the business forms are a few with insignificant improvement in dollars. Taking a specialized jump, the conclusion to end client benefit satisfaction can be accomplished by changing over the capacities offered by the BSS and OSS layers by administrating and demonstrating the information reliably. SOA Item Sellers offer very adaptable administration conveyance stage for BSS and OSS layer capacities. Some of the important key benefits are listed below.

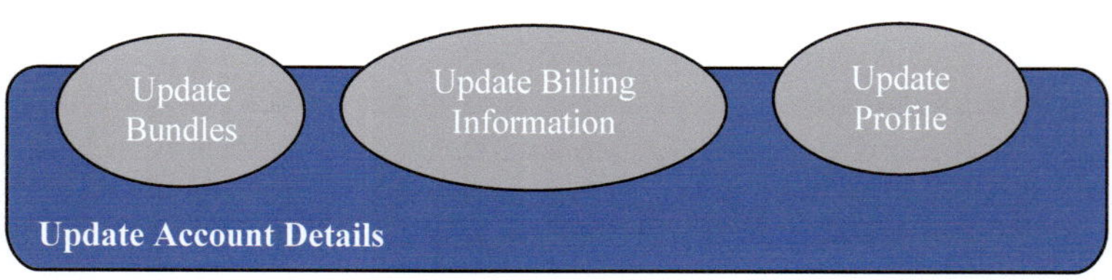

Fig. 6.10 Business process

1. **Cost Reduction**

At the point when the web channels begin to give expanded usefulness and engage the endorser of deal with their records and membership all alone in a client benevolent condition, the calls to the CSRs will diminish. These will help in call centre enhancement. Once the establishment squares have worked as administrations in view of SOA Standards, these administrations can be utilized to execute another business procedure or revise a current one without critical advancement costs.

2. **Greater Customer Satisfaction**

In this era, the service operator is capable to provide a one-step process to sign up in one or more offerings by designing converged business process. Although this might be a case in various operators, SOA facilitates easier IT management from an operator's viewpoint. The user understanding through the web channel can be considerably raised with user-friendly navigation controls and high-end graphics with the use of technologies based on RIA.

3. **Shorter Time to Market**

As shown in the previous diagram, revising a business procedure or including new ones is generally a more straightforward errand, which can be accomplished with noteworthy reuse of IT resources at an administration level. This fits shorter improvement cycles for upgrades and changes. With the business forms being demonstrated as normal to the different channels such as Web, CSR, Merchants and so on, an adjustment in a current business process or another one can be presented at the same time over the channels.

Figure 6.11 shows some of the common business processes and their services which can be modelled based on SOA.

6.2.1 Service Hierarchy

The significant role of SOA infrastructure and its capability can be identified with the help of SOA Infrastructure. It is the most important aspect of IT approach which spans the entire project. IT projects are designed to fit within an SOA and are for that reason, just a section of the larger SOA initiative. The SOA infrastructure projects are different than typical SOA project to some extent. In the infrastructure SOA projects, at delivery time, the main focus is given on the infrastructure products instead of providing new business services [10].

SOA infrastructure communications keep the focus on that it has to enable the delivery teams and then supply SOA projects more speedily. SOA also needs to make sure that overall SOA is convenient. Realization of SOA has two important ways as SOA plays a vital role in both runtime and design time activities. SOA application must follow guidelines of reference architecture and procedures while development. The application can be considered as SOA application when it has followed these

BUSINESS PROCESS	SERVICE
Customer Registration →	Registration Eligibility Check. Profile Check. Customer Credit Check. Creating Customer Account. Creating Customer Profile Update Customer Tariff / Talk Plans. Activate Customer Account or Services. Update Customer Contract.
Account Maintenance →	Update Customer Account Details. Update Profile. Update Customer Billing Details.
Subscription Maintenance→	Update Product. Change Customer Telephone Number. Update Customer Talk Plan / Tariff. Activate Services. SIM Swap. Contract Renewals. Disconnection. Transfer Subscription.
Upgrades→	Check Upgrade Eligibility. Calculate Fees. Provide Upgrade Options. Activate Service. Fraud Check.

Fig. 6.11 Common business processes and their services which is based on SOA

rules. The SOA team, i.e. EA team has the responsibility of implementing application and services as per guidelines of SOA reference architecture [1]. The development of application and services as per the planning and strategy of SOA architecture places demands on the SOA infrastructure. These demands can design time as well as runtime. The former includes the composition of application and service, the orchestration of the services, loose coupling, controlled service discovery, service versioning, management of security policies and data format conversions. The success of SOA success depends on the capability of the development team to fully preserve and control the situation as the SOA matures on over its time. These teams required infrastructure to offer a hand in observing and organizing the SOA situation. Applications that can self-possessed, principally of services, which are often in a hierarchy are the service-oriented applications.

Various levels in the SOA are depicted in Fig. 6.12. The topmost level is the most important level and it includes one or more combinations of services. In these, each of service controls a flow of activities, such as processing an applicant's application for assurance of exposure [8]. One or many business services are invoked by every combination of services. The next level, i.e. the second level, is comprised of services which look after relatively low-level business task. For example, incorporation service, this service invokes a sequence of business services to approve or authenticate the details that were delivered by an insurance-policy representative. The incorporation service invokes another business service in the case where business services

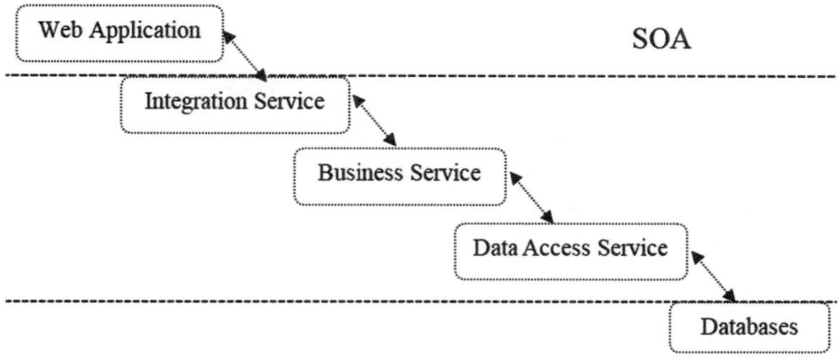

Fig. 6.12 SOA levels

return values that are critical [8]. Then, the business service calculates a quote and returns the quote to the software, such as a web application, that invoked the service-oriented application. The third, i.e. lower level, contains data-access services. These services are intended to read and write from/to data storages like databases and message queues [11]. Business layer invokes a data-access service, however, easy access to these services may cause dissimilar usage of services. E.g. if the web application has easy access to data-access service, then it can easily reassign initial values for the parameters in the form.

The most vital feature of SOA architecture is flexibility. Various forms of incorporation services offer facilities to different requesters to invoke other incorporation services. This feature provides strength to the requester to access different kinds of services from within a service-oriented application. Concisely, we can say that the requester gets access to additional services at a single position and, on the other hand, is capable to offer business services to users an alternative way.

The communication between mobile phone devices of telecommunication's BSS or OSS systems and core carrier networks is referred to as service provisioning. It contains the establishment and consequent modification of payment on the network. Typically, these types of events are mobile handset activations. These modifications may be improved with the help of activation of new services or deactivation of existing services which need a network.

The end-users can get the benefit of high-end and innovative services if the network can provide high speed and high bandwidth networks. However, provisioning new service may face various challenges like market situations; it may take a longer time to up in the market and sky-scraping IT accomplishment cost. In the telecommunication industries, there is a need for emerging models that combine the service and contents to provide a wide range of capabilities in the network, e.g. Triple and Quad Plays. The service provisioning is flexible, highly available and doesn't cost much. The SOA business process includes the various services like customer management (CM), services from a variety of BSS, OSS systems and network fundamentals like billing as well activation and supply chain management (SCM). Through the automation of the manual intervention required for provisioning and activation, new service offerings across the spectrum (wireless, fixed, broadband, media service, etc.) can be provisioned and activated without additional steps.

Additionally, it has the ability to deliver services at lower functioning costs and shorter time period to enhance the market. The architecting and the provisioning process based on the SOA principle will help as proofs for shipping and as the drive for IP-based principles. Clearly, an IP-based infrastructure will enable the flawless transition of service across mobile, broadband, fixed lines and host of new applied services based on the delivered content [12].

6.2.2 *Execution of Call Flow*

The execution of call will first establish a connection link for communication between mobile devices and wireless network with a dedicated link to achieve the task. It will support all enterprises in the smooth execution of dedicated work irrespective of the communication link. Any service that performs a specific task is a self-sufficient part of the software. A call is a full broadcast event requests between participants. Once participants are connected, a medium flow is established between them. The communication can be achieved through many channels at the bottom, for example, audio, video or other channels.

6.3 Context-Aware Business Model

Generally, any business model helps the organization to identify defects and improve the quality of the business services for achieving dedicated goals. The network-related issues have to be identified and steps have to be taken to overcome it with available techniques and technology for future enhancement and hence context-aware business model will help to achieve context awareness which is a positive feature of mobile devices paired with location alertness. Location, as we know, may be difficult to find out and apply it to operate the devices based on the region. Context, on the other hand, may be identified and applied with more flexibility to many mobile users, principally with dedicated smartphones. It is a known fact that context services originated for the purpose of realizing computing everywhere and anytime, which requires dealing with the changes in connections in the available environment with standing computer systems. The way of context computing is also applied to contextual type application design and many support small as well as large business process management issues.

The context-aware computing service is particularly defined as a style of computing which can be used to work with situational and ecological/environmental information regarding people, places and things. They are collected with a motive to understand the immediate needs and proactively suggest solutions, provide situation-aware and utilizable content, functions and experiences. It is expected that by the end of next year, more than half of the Fortune 500 companies will have context-aware

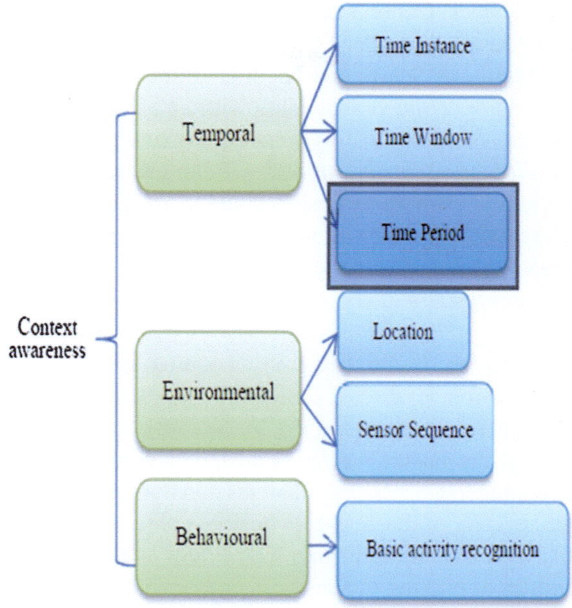

Fig. 6.13 Business processes and context

computing initiatives with their own decisions about existing system disadvantages and security concerns for business application and achievement with dedicated computing solutions.

Any start-up or already settled business process models have been ranked or rated on the basis of the top three strategic technologies suggested by the context-aware technology; it has been extensively accepted as a capable driver for next-generation technology evolution and future enhancement. A better understanding of the background of the context-aware business process management will help to incorporate its capability to achieve its dedication and enlarge the scale for future running application in the networking areas.

Context and Context Awareness

Industries required to capture contextual information and adapt the service models to cope up with changing business demands. This process of adopting changes should be fast as the popularity of business increasing rapidly. Dey et al. defined context as '*any information that characterizes a situation related to the interaction between humans, applications and the surrounding environment*' [13]. However, context is ubiquitous and difficult to handle due to the complexity of information selection, distribution, modification, and utilization. Business process and context are shown in Fig. 6.13.

The performance of the business process can be estimated based on the choice of context information used by business applications. Context is a silent but extensive performer in all interaction and communication process that can outline human understanding and conversely can be manipulated through a lens of applications and devices. Context awareness is not a new topic and it can be generally found in other research models. There are different definitions of the context awareness in the literature. Schilit and Theimer defined it as '*location awareness*'. Dey et al. have defined the context awareness with *an emphasis on its usefulness in identifying the entire human-application interaction situatio*n [13]. In spite of the different definitions of the context awareness, we can say that it consists of all applications and users including even an extensive range of information identification, adoption, analysis and modelling. The various parameters are taken into consideration while identifying meaningful and useful contextual information, in the recent context-aware system, chronological and ecological parameters are used to recognize contextual information.

However, traditionally, in designing business process methods, the contextual information is rarely considered; hence, it cannot take advantage of the context awareness. There is a need for adaptive systems that can integrate context-aware ability into business process design as nowadays attention on internal and external context-related business changes and dynamic workflow patterns are increased. The perspective of the design of business process management is changed from activity-oriented modelling to rule-based item/artefact-oriented. That shows massive capabilities of making the business process more responsive, flexible, agile and adaptive to changes.

6.3.1 User Perspective

Users can communicate with each other with the help of context-aware models, in order to achieve dedicated tasks. Sometimes users and applications communicate in the form of context-sensitive manner.

Specifically, we think about communication with respect to some available tools like IM, SMS, Email as well as telephones. Many inbuilt features are provided by tools from a user's perspective of communication. On a primary level, context-aware business model from a user's point of view can be used in M-commerce for location and tracking services, mobile services, monetary services, shopping as well as advertisement, contents and tracking services, mobile financial services, telecommuting, mobile phone games, reimbursement systems, entertaining and location-sensitive satisfied content services. The future services from a user's perspective for business models can be used to grow his business for all different services which are based on context-aware business model specially designed to enterprise businesses.

- **Perceived value for the service to users**: It is implemented based on utility and communications in which content should be integrated, relevant and well known. It includes personal and user-generated content.
- **Perceived easiness of use to users**: It is based on clear/understandable services and ease of direction finding on small present screen and simplicity of interaction with the mobile service without even needing to put a lot of effort on personalization.

Scalability to satisfy the necessities of dissimilar mobile devices and its networks is a very important aspect for users and services.

- **User trust**: is as important as mobile services. It became more user centred and user controlled with respect to privacy.

6.3.2 Network Operator Perspective

To improve operations from a future perspective as compared to existing systems of communication, using the current set-up of network communication offers many best features but lacks some significant improvement towards the future enhancement of communication. But context information has been changing dynamically and the need to keep information confidential the operator has to establish control over it. Hence, context awareness plays an important role in handling such types of problems occurring in Android and other platforms.

The actual realization of context awareness for the purpose of the intelligent and controlled solution by the network operator comes when users are naturally interested in protecting their privacy which requires the information to be collected for the purpose of the context-aware operation.

The network operator has many capabilities to access meaningful data. It may provide a centralized as well as a dynamic solution for information management as well as identify devices and provide context-aware services. The proposed method for the addition of new capabilities on the network and communication links between the IoE components, fresh mode and its types of community context are promising to the users. Additionally, the types of environment, user profile, content data, device ability (e.g. storage battery, CPU, wireless), user nearness to other users and the overall quality of the available communication links to other devices and applications are fundamentals of future context.

The main role of network operator is to enhance the capability of the network with performance with the help of the context-aware system as compared to the existing network system.

6.3.3 Third-party Perspective

It is the way of representing different processes of enterprises so that we can analyse the current process and improve it for future enhancement. It has been helpful for leading the growth of the business. From a business point of view, we need more flexibility and adaptability. The service is generally categorized for third-parties as follows.

1. The service provider party
2. The service consumer party

Service may be requested by a third party or an external entity, and the service provided is a public service.

Example: Amazon and Google already provide external services.

Considering the social sites for different services, they completely depend upon essential security concerns provided by the service provider and its respect to use that service to achieve the dedicated task for external entity authentication. Context-aware access services are dedicated to future growth and accommodating the existing system to accept changes for future growth. It completely depends upon the services which we use from a third party or external entity for our requirement in order to accomplish the dedicated task. User context awareness helps to increase user assurance for their endeavours such as a mobile application where they deliver with the help of special user personalization and user suggestion by analysing various data sources through several process workflows.

6.4 Conclusions

This chapter has given us an in-depth understanding of context-aware systems and its usage in the telecommunication industry. With the emerging applications and advancements in the Internet of things (IoT), context-aware applications play an important role in improving user experience and enhancing the quality of communication and presentation. This chapter has also described the layered architecture of context-aware systems along with its applications. Tied to this, the importance of SOA and its establishment in realizing the telecom industry applications are also dealt with in this chapter. On the whole, with the rapid changes that are happening in the business process management, changing requirements the context-aware systems tie with SOA plays an important role in helping processes adapt to the changing models.

References

1. Dustdar, S., Jørstad, I., Thanh, D.V.: A service-oriented architecture framework for collaborative services. In: Proceedings of the 14th IEEE International Workshops on Enabling Technologies: Infrastructure for Collaborative Enterprise (WETICE'05), (2005)
2. Sur, A., Arsanjani, A., Ramanthan, S.: SOA based context aware services infrastructure. In: 2007 IEEE International Conference on Services Computing (SCC 2007)
3. Chihani, B., Bertin, E., Jeanne,F., Crespi, N.: Context-aware systems: a case study. In: The International Conference on Digital Information and Communication Technology and its Applications, June 2011, Dijon, France
4. Jones, K.: Building a Context-Aware Service Architecture (2008, Dec 12)
5. Ramanathan, S., Arsanjani, A., Allam, A.: Leveraging context-aware services to transform evolving telecom industry business model. In: 2007 IEEE International Conference on Service Computing (SCC-2007)
6. EMC: Content services in telecommunications. EMC White Paper (September 2005)

7. Smith, J., Kealy, A., Williamson, I.: Location Based Services—The Underlying Technology
8. Service Oriented Hierarchy. https://www.ibm.com/support/knowledgecenter/da/SSMQ79_9.1.1.1/com.ibm.egl.pg.doc/topics/pegl_serv_overview.html
9. Harris, T.: Service Oriented Architecture Telecom. https://www.thbs.com/downloads/SOA-for-Telecom.pdf
10. Selamat, M.H., Al Kharusi, A.: Service oriented architecture in education sector. IJCSNS Int. J. Comput. Sci. Netw. Secur. **9**(5) (2009)
11. Schilit, B.N., Hilbert, D.M., Trevor, J.: Context-Aware Telecommunication Services
12. Chihani, B., Bertin, E., Suprapto, I.S., Zimmermann, J., Crespi, N.: Enhancing existing communication services with context awareness. J. Comput. Netw. Commun. **2012**(493261), 10 p (Hindawi Publishing Corporation)
13. Dey, A.K., Abowd, G.D.: A conceptual framework and a toolkit for supporting the rapid prototyping of context-aware applications

Chapter 7
Security Issues in Context-Aware Systems

Abstract In the era of ubiquitous computing, security issues in context-aware pervasive systems and applications play a crucial role. Security, privacy and trust are the key factors to support secure information access. This chapter presents the motivation towards the need for security solutions to context-aware systems and discusses risk assessment methodology. This chapter also presents common terminologies of various security features and comprehensive state the art in order to understand the potential areas which need further research. Threat analysis and attack modelling are important tasks to design security solutions. In the sequel, the next part of this chapter presents activity modelling and threat taxonomy for context-aware systems. The relationship between machine learning, ontology learning and personalization is also discussed in this chapter. Finally, this chapter concludes with an integrated framework for security, privacy and trust and by proposing a security algorithm.

7.1 Motivation

To extend the information-rich services, smart spaces sense contextual and situational information for adaptive context management. Smart spaces available in smartphones and in the surrounding are continuously and proactively sense and react to the contextual information. The objective of this sensing is to meet the user's preferences and expectations. Ubiquitous computing paradigm, distributed operations and architecture add several threats and vulnerabilities in context-aware pervasive systems and applications. Cyber attackers and hackers are more potential entities and take maximum benefits of mobility and distributed access. Context-aware pervasive systems are prone to enormous risks which include threat, attacks and hacks and need to be investigated to avoid the loss.

Before we design a security solution to context-aware systems, it is important to perform risk evaluation and assessment. It is a procedure of classifying and estimating various risk factors and their corresponding consequences. It is a process of identifying and evaluating various risk factors and their corresponding consequences [1]. According to NIST [2], risk assessment also includes internal and external threats, internal and external vulnerabilities, impact and its likelihood, i.e. the degree of harm.

P. N. Mahalle and P. S. Dhotre, *Context-Aware Pervasive Systems and Applications*, Intelligent Systems Reference Library 169,
https://doi.org/10.1007/978-981-32-9952-8_7

It is also equally important to analyze existing security solutions and their effectiveness as well as the areas that need to be addressed. The risk assessment process includes various phases like context establishment, risk assessment, risk treatment, risk acceptance and risk communication/monitoring. A generic risk assessment process is presented and discussed well in [3] and depicted in Fig. 7.1.

For context-aware pervasive systems, trusted and secure interaction between various devices and sensors is very important. With the increasing number of smart spaces acting as content providers and actuators as well as their functionality, information exchange has become a crucial point. There must be a safe and trusted way of swapping information with and between things and sensors that do not imply a heavily administrated environment but rather provide a good compromise between scalable technologies, privacy and safety of the information. The security involved must scale in number, supporting many devices interacting and adjust to the available resources, since 'things' might not have a lot of available resources (e.g. sensors). In the sequel, there is a need for a generic framework that enables trusted context and actions to be exchanged securely and with the required privacy constraints. There is a need for a set of protocols that will exchange the information based on identity and deliver actions to devices from smart spaces. There should also be a provision for the exchange of secure and certified identity dependent context and actions across multiple devices and identities. The detailed study should be initiated to analyze the

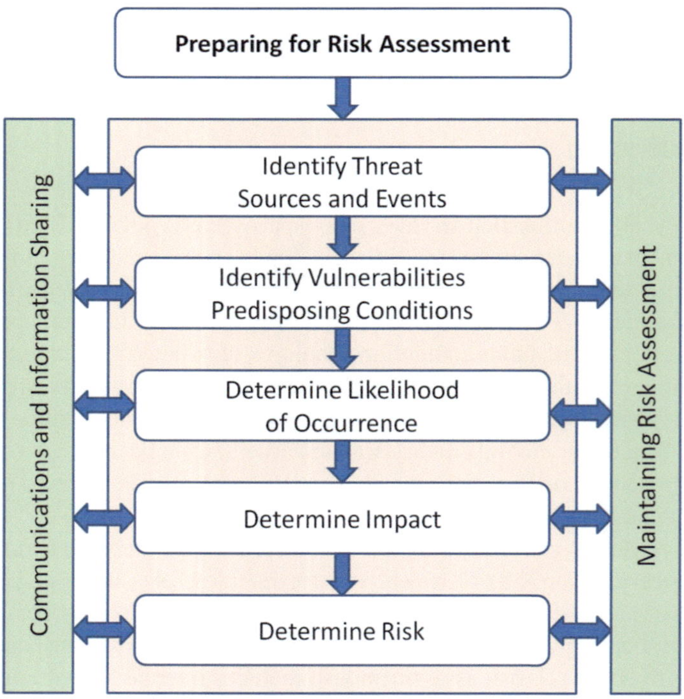

Fig. 7.1 Risk assessment methodology

security capabilities of IPsec, Android application development platform and other mobile application development technologies. Another objective is to define different security levels that accompany and bolster the space definition by introducing new concepts such as using APIs for defining security level using ambient awareness or using the better-than-nothing security (BTNS) to define best-case scenarios for interacting with things.

In addition to this, for secure and distributed handling of sensitive information, there is a need for privacy-enhancing technologies, augmented trust and auditing with new layers of assurance. Information about user and devices is distributed across different points in the network, and often they leave trails that lead back to privacy-threatening information. This creates a need for the user to be able to track what is common knowledge shared about him on the network, and what type of information is spread throughout the different points. This implies that there are tools that allow the user to perform the auditing of the distributed information, while keeping the entire system manageable, regardless of the complexity of how the information is scattered. In the sequel, there is a scope to design framework that allows handling distributed information and auditing with different levels of assurance about the privacy of such information pertaining to the user. There is a need for mechanisms and protocols to investigate distributed data, specially geared at secure and private distributed data management. Researchers should also evaluate and propose tools for auditing and visualizing distributed data, focusing particularly on user personal data auditing and user-centric enhanced auditing visualization.

For every context-aware system, there is a need of dedicated context-aware security manager (CASM) controlling the requests coming in and data going out of devices. CASM is responsible to perform actions like authentication, authorization, access control, privacy enforcement (e.g. non-disclosure or selective disclosure, anonymity and confidentiality) and accounting. To support the above-mentioned functionalities, secure context management framework is mandatory to separate the set of applications from context sources. This helps to share the context among applications and allow applications to flawlessly work in different smart spaces and environment. In addition to this, there are limitations of performance parameters like network bandwidth and battery power (energy) and the communication overhead should be optimised. The major responsibilities of this framework consist of the fetching, handling, storing, swapping and providing context information. The high-level view of this framework and the functional component interacting with it is depicted in Fig. 7.2. Applications and services can be written using a uniform and standardized interface to access context information from multiple sensory and data sources across the web. At the same time, data sources can be written against another standardized interface to provide context data. This framework takes care of everything in between, including fetching, handling, distributing and controlling access to available context information.

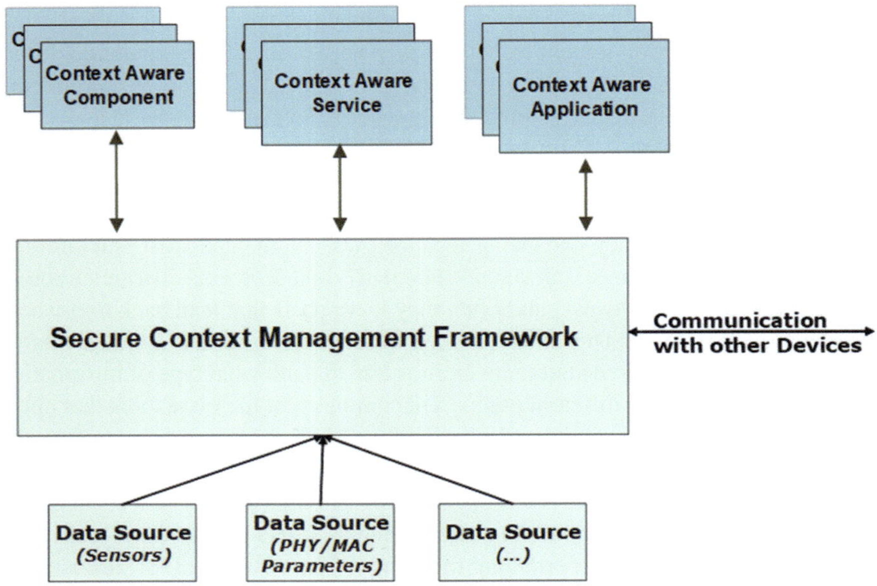

Fig. 7.2 High-level view of secure context management framework

7.2 State of the Art

Security is a vast administrative domain, and there are various security features. In the literature, these features are defined in different perspectives. However, using common terminology and agreeing about the definitions is important to understand each other. These definitions are presented in Table 7.1.

There are many potential research projects which have been carried out to adders the security issues in context-aware systems. In [4], the context-aware usage control model is presented and discussed. The proposed work also extends traditional UCON [5] model. The key components include rights, authorisations, access control and obligations. However, the formal specification and implementation details are completely missing in the proposed work. In [6], adaptive access control scheme using context awareness is discussed. In the proposed work, context is integrated with access control. The proposed work is not validated against known attacks, and the efficiency of the proposed algorithm is not evaluated. Security and trust issues in context-aware applications are nicely explained in [7]. Various challenges and the proposed solution with respect to the design of context-aware systems are well discussed in [7]. The motivation behind the security requirement for context-aware applications is introduced in [8]. This work also presents security attacks and respective mitigation techniques. Analysis of existing security protocols is also discussed in [8]. However, it lacks the algorithms and its implementation. Predictable information security-based context-aware trust model and its statistical analysis in the presence of suspicious entities and authorities are presented in [9]. Review of the

Table 7.1 Definition of security terminologies

Sr. No.	Term	Definition
1	Anonymity	Anonymity means hiding the identity of an entity involved in the communication process. A popular approach is to replace the identity by a pseudonym. If the same pseudonym is used repeatedly within the same context, this identity can build up a reputation (a.k.a. Pseudonymity). If the pseudonym is used for a short time only, it is very hard to get a grip on this identity and thus the user remains anonymous
2	Authentication	In principle, one can distinguish between two different kinds of authentication. Entity authentication denotes methods that mutual identity establishment, thereby helps to prevent impersonation. Data authentication, on the other hand, means the corroboration of the data's source. With interacting computer systems, authentication is always the latter (i.e. data authentication)
3	Authorisation	The authorisation is about enabling an entity to access a resource based on rules (e.g. an access control list) configured for the (authenticated) identity of the entity. This can take several forms (performing a requested operation or failing it, providing access keys or Kerberos certificates, etc.)
4	Auditing	The process of checking logs and records of events for anomalies like attacks or system faults
5	Confidentiality	Ensuring that information is not accessible by unauthorized entities
6	Data Freshness	For certain data, e.g. credentials or medical measurements, their freshness should be guaranteed to avoid replay attacks This means that an adversary cannot simply reuse old data in subsequent communication
7	Identity	Identity is either a name that identifies a person or an object or an abstract concept using an 'identifier' (e.g. a string of characters like a name). For example, each node of the network must be linked to a unique identity
8	Integrity	The property that data has not been altered or destroyed in an unauthorized manner without being detected by authorised parties

(continued)

Table 7.1 (continued)

Sr. No.	Term	Definition
9	Intrusion Detection (System)	Intrusion detection involves the monitoring of events that take place in a system or a network and the analysis of these events for the detection of malicious activity and attacks. These attacks are generated by people from the Internet, from authorized users who are trying to gain access to even more resources than they have been assigned to and from authorized users who are taking advantage of the privileges they possess. The term intrusion detection systems is related to software and hardware products automating the above procedure
10	Key	A special and unique sequence of bits, which has particular properties to be suitable for encrypting data or verifying identity using cryptographic algorithms
11	Key establishment	Key establishment is any process where the key is exchanged securely between two parties and becomes available to communicating parties for succeeding cryptographic use. Thereby, a potential adversary must not be able to derive the secret key. Key establishment is a mandatory step for many cryptographic mechanisms as encryption or authentication
12	Non-repudiation	Non-repudiation is given if the sender of some data cannot successfully deny afterwards to have sent the data
13	Reputation	A way to build up trust based on previous experience (could be transitive, reputation reported by trusted others). In computer systems, reputation can be established by comparing the observed behaviour with the expected behaviour and storing the result for future reference. In addition, trusted systems may provide second-hand reputation information
14	Trust	Trust is a human concept, which is used to indicate a feeling that the person or object being trusted will not display unexpected behaviour. In the context of networks, the trust of a node A is expressed by a node B when it provides A access to certain services or information without further authorization. That is, the trust B has in A is represented by the level of access B grants to A. Consequently, this trust can be increased, limited, revoked or forwarded
15	Security Association	A mutually established state containing the means to authenticate each other and to communicate directly using a secure channel

security-related issues in the context-aware system is nicely presented and discussed in [10] but the proposed work is missing. Context-aware user authentication supporting proximity-based login in pervasive computing is introduced in [11] where user authentication based on the context is proposed and discussed. However, the verification of the proposed work against attack is not presented.

7.3 Activity Modelling and Threat Taxonomy

In order to evaluate the security of the system, we must analyze the threats and vulnerabilities that are part of the functionalities and the environment. Obviously, both are far from fixed as the user may add or remove devices from the network and being mobile, the user keeps experiencing different environments. One way of analyzing security is by using attack trees. An attack tree is a method of mapping out security risks by putting the attacks on a system in a tree form. Attack trees provide a formal, systematic way of defining the security of systems, based on multiple attacks possible on the system. Attack trees are basically the general trees with many levels which include one specially designated node called root, terminal nodes which are called as leaves and internal nodes which are called as children. To deal with the specific attacks, the underlined attack tree and its structure become more complex; complete attack tree consists of thousands of nodes and more than thousands of different paths which are all leading to the termination of attacks towards leaf nodes or terminal nodes. Attack trees are very useful for determining what threats exist and also provide mitigation technique to deal with these attacks. Attack trees are the bottom-up process, at first, the child node conditions have to be satisfied to make the parent leaf true, and when the root conditions are satisfied, the attacks are complete. Each tree can be satisfied by only its direct child node. In some cases, a leaf may be a child of another leaf which uses multiple steps in order to carry out an attack. In other words, it represents attacks available in the system in a tree structure, with the objective to be represented by the root node and different ways of mitigating these objectives as leaf nodes. The node of an attack tree can be decomposed in one of the two ways: a set of attack sub leaves, where all must be satisfied for an attack to succeed, and it is represented as an AND decomposition. A set of attack sub leaves, where anyone should satisfy for an attack to succeed, these are represented as an OR decomposition. Attack trees can be represented graphically, the AND decomposition is represented in Fig. 7.3.

The AND decomposition represents a root R1 that can be satisfied if the attacker satisfies each of Leaf 1 and Leaf 2. The OR decomposition represents a root R1 that can be satisfied if the attacker achieves any of the Leaf 1 or Leaf 2. Fundamentally, in attack tree, there can be any combination of AND and OR decomposition. As discussed earlier, leaf nodes are used to terminate the attack. OR decomposition is used when there is a need to generate new scenarios, wherein AND decomposition is used when there is a need to extend existing scenarios. Internal or intermediate

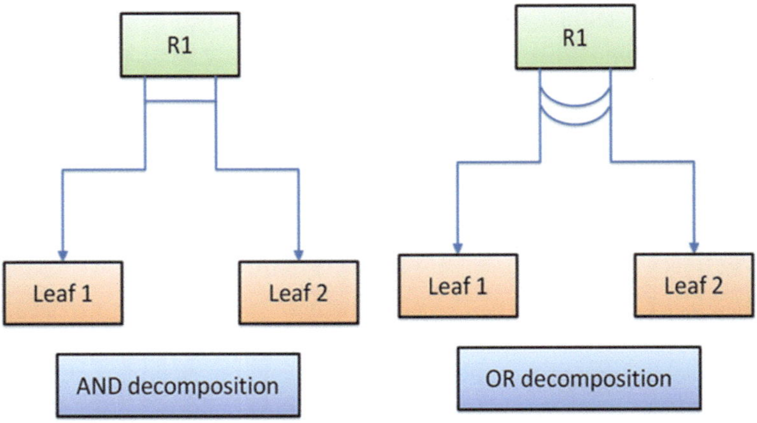

Fig. 7.3 Overview of attack trees

nodes never appear in the attack scenarios as they are always elaborated by the goals at next lower level.

Context-aware pervasive systems are a convergence of operational technology (OT) and information technology (IT). The objective is to gather contextual information from the sensory data and smart spaces. There are raising threats due to Internet connectivity which is a public network, and there is a need for preventive maintenance rather than reactive maintenance. In order to protect the system, it is essential to analyze all the components and possible attacks on them. This attack analysis is presented in Table 7.2.

7.4 Mitigation

First of all, security solutions should be specified in a way that makes it clear which threats the design should protect against. The likelihood of an attack is usually determined by the amount of profit the attacker can gain balanced against the cost and risk to execute a successful attack. So the goals are to reduce the value of the loot after the attack, make the attack more expensive and risky to perform and also to reduce the amount of information leaked about the potential value to be gained by the attacker. And when an attack is successful, we should gather as much evidence as possible to find and convict the attacker when (or if) he or she is caught. All security solutions should take into account that the human user is most likely the weakest link in the chain. This implies that the automated security solutions should encourage the user to act securely while not getting in the way too much.

Like software, there is also a software development life cycle (SDLC) for context-aware systems and applications. There are certain recommended security measures and task to be performed during the software development life cycle of context-

Table 7.2 Attack analysis

Technology	Component	Possible attacks
IT	Data Analytics, Business Applications and Mobile Devices	DoS, Side-channel attacks, Cloud malware Injection, Authentication Attacks, Man-in-the-Middle, Mobile device attacks
	Office Applications, Mail and Web Services	Phishing, SQL Injection, Malware, DNS poisoning, Remote code execution, Brute force attacks, Web Application Attacks
Demilitarized components		
OT	SCADA, Other Controllers and Operators	IP Spoofing, Data Sniffing, Data Manipulation, Malware
	Distributed Controllers and Management System	Replay attack, Man-in-the-Middle Attack, Sniffing, Wireless Device Attacks, Brute force Password Guessing
	Sensors, Motors, Actuators, Transmitters, Embedded Devices	Reverse Engineering, Malware, Injecting Crafted Packets or Inputs, Eavesdropping, Brute force search attacks

aware pervasive systems and applications. In the requirement gathering phase, attack surface analysis, threat analysis and modelling are recommended in order to minimize the security risks. When the system is to be designed, more emphasis should be given on designing secure architecture using multiple test cases. During the coding phase, secure coding guidelines are recommended to prepare static analysis. Functional and security penetration testing is required to be performed during the testing phase. During the release or deployment phase, antivirus continuous inspection is required along with the installation of the latest patches.

In the view of the security feature of context-aware systems, access control is one of the important functionality. Access control rules are designed and formulated for controlling access to information in order to decide which object/subject is allowed to access what object/subject. Role-based access control, access control list and policies are the variants of these rules. Interoperability is a major concern when standard-based policy language was not available. The proposal of XACML [12] put forth by OASIS is the first proposal which enabled security solution providers to use standard-based languages to define the set of policies and to offer added standard solutions. Although XACML was the starting point towards the definition of standard policies, it is only focused on one kind of policy, the resource access control ones. More or less at the same time, another kind of policies emerged to cover specific aspects for identity management, for example, P3P [13], is used to define online privacy announcement information policies between users and services. Current systems have incorporated in some way these kinds of standard policies, for example, Shibboleth [14] and Liberty Alliance [15] providing a definition of access control policies by means of XACML, or PRIME making use of P3P. But there are other types of policies, described below, that need to be defined in a standard way in the next generation of policy-driven systems when distributed scenarios in the context-aware domain are considered. In view of this, one of the approaches is to investigate policy-driven approach [16] which includes privacy release policy, identity management policy, resource access policy and accounting policy.

To address various security issues in context-aware pervasive system and applications, there is a need of dedicated module which provides advanced security options like authentication, authorization, access control, privacy enforcement, trust management and accounting. The security, privacy and trust module is a logical component dedicated for defining security, privacy and trust, and their decisions are based on contextual information provided for the environment and requested services. The assumption here is that the security component has access to contextual information. These components perform adaptive security management which is used to ensure interactions which are context-aware. Integration of different levels of security, privacy/anonymity and trust ensures adaptability and flexibility.

Integrated Framework

An integrated framework for providing mitigation to various threats and attack is presented in Fig. 7.4.

The functionality of the security agent is to ensure data confidentiality and data integrity with respect to the mapped profile. It also assures data origin authentication and the output is given to trust agent. The security needs of the communication phenomenon are varied. Mainly, the security needs depend on the type of communication and change management in a context which includes location change or exchange and transfer of sensitive user data. The security agent is responsible for compliance

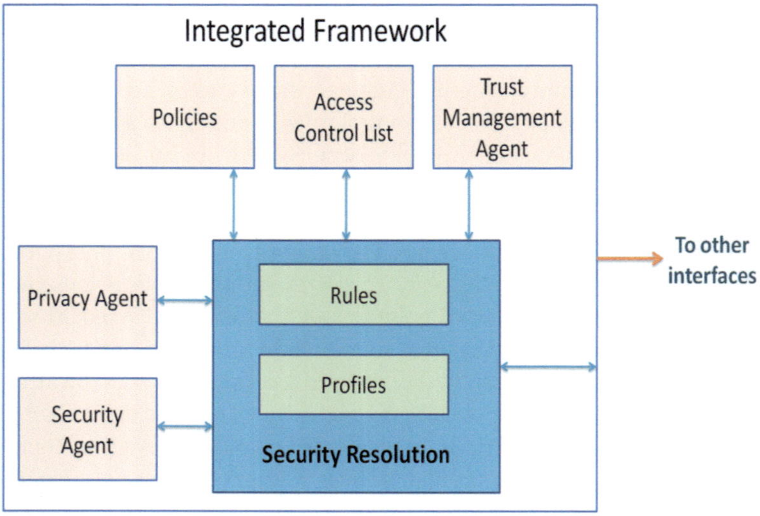

Fig. 7.4 Integrated framework for security, privacy and trust

of security mechanism. This security agent is a software entity which is accountable for finalizing and assigning the required security level.

There are three possible levels of security for communication—low, medium and high which are defined below:

Low—delivers very common services and allows the exchange or transfer of trivial and common data.

Medium—this level requires general protection for the communication even the data which is being exchanged is trivial and common data. This level includes common transaction and exchange in daily life.

High—provides highly secure access to the services and involves transfer or exchange of credentials and data which is more sensitive and important. As we know that the security always comes with the cost, the provisioning of high-security level will surely affect the network performance and utilization.

There are multiple rules, profiles and types of context information available in the communication. These combinations of these rules, profiles and context types are responsible for finalizing the security level. The requirements of underlined applications and services are also considered in addition to the above three factors, and this consideration is indirect via the application/service profile. Trust and trust score play a very important role to protect data and device credentials from illegal and compromised objects. The factors like identity management, knowledge, experience, recommendation and roles are responsible for trust, trust score and trust establishment. Key security features like authentication and access control and their respective mechanisms are also prerequisites and inputs for trust management scheme.

The trust levels are defined as follows:

Unknown—entities which do not have any information about knowledge, experience and recommendation initiating communication for the first time.

Untrusted—entities for which the access is strictly prohibited even the access was granted in the past. The possible reasons for this are either the rights are revoked or entities are completely untrusted.

Trusted—entities that have a good trust score among them and the data transfer already took place between them with the shared trusted key.

Generally, privacy is considered as a security feature which protects the identity of sender and receiver thus ensuring confidentiality. Supporting parameters like nature of traffic, multiple patterns involved in the traffic, traffic size, location, time and the message size are also the source of information for the communication. Privacy violation can take place by device tracking, user identification and forced data access. Aspects, like maintaining information privacy, preserving the anonymity of the users for distinct scenarios and maintaining location privacy of a node, should be considered to deal with the privacy issue.

Policy and policy-based clearance play an important role in context-aware applications and mainly done by security resolution point. The policy is a container where rules are defined along with the set of respective defined values, parameter range which is available in the security profiles. Security resolution point includes all these

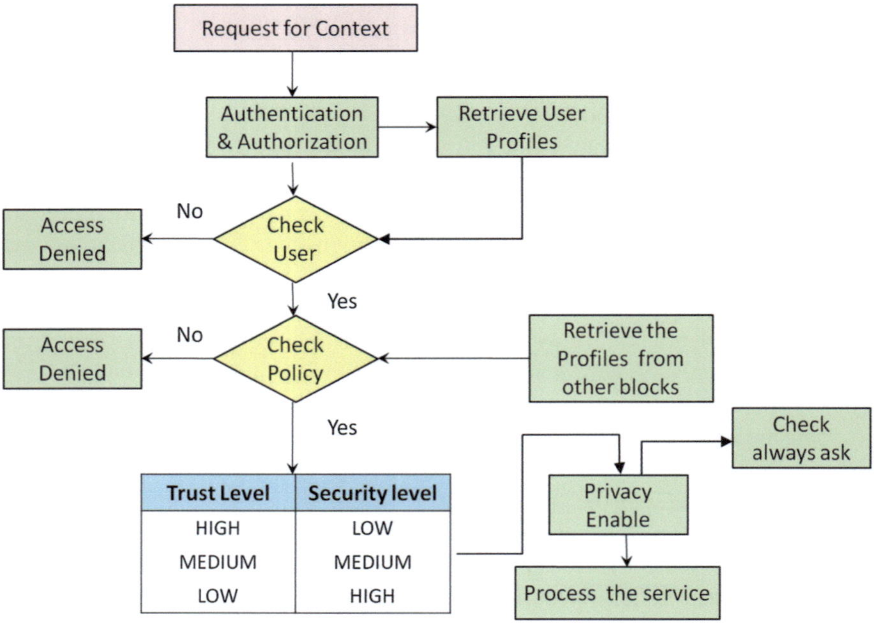

Fig. 7.5 Proposed security algorithm

rules required for finalizing policy decisions. The authenticated request is always governed by these policies to confirm secure access control. At the security resolution point, user profiles, services are taken into consideration and in turn, modification, updating and deletion of the user profiles are initiated.

In the view of the integrated framework presented in Fig. 7.4, for security, privacy and trust, the proposed security algorithm is depicted in Fig. 7.5.

7.5 Conclusions

Whenever security is designed and then implemented, the question remains (and should remain) whether we covered all the (realistic) threats in our design and implementation. In complex systems, such as context-aware system and even more so in ubiquitous applications, it is not possible to know all the threats and vulnerabilities in the system. Both in the design and implementation phase, vulnerabilities may be overlooked or introduced that are not covered by a security measure. For this reason, threat analyses should be conducted in all phases of development and the lifetime of context-aware systems and applications. Nevertheless, this chapter gives an overview of activity modelling and threat taxonomy along with mitigation techniques in preliminary form for context-aware systems. Further work is still needed to refine and standardize the security architecture for context-aware systems.

References

1. Definition of Risk Management: ISACA. http://www.isaca.org/Pages/Glossary.aspx?tid=774&char=R
2. ISO/IEC 27005:2011. http://www.iso.org/iso/catalogue_detail?csnumber=56742
3. The NIST SP800-30: Guide for Conducting Risk Assessments; by the NIST Computer Security Division (CSD); Draft version (Revision 1), September 2011, File Name: SP800-30-Rev1-ipd.pdf: Pages.: Appendix B, pg. B-8; Chapter 1, pg. 1; Pg vii; Chapter 2, pg. 7 to 13; Figure 5—Risk Assessment Process on pg. 19 (http://csrc.nist.gov/publications/drafts/800-30-rev1/SP800-30-Rev1-ipd.pdf)
4. Almuatiri, A, Siewe, F.: CA-UCON: a context-aware usage control model. In: 5th ACM International Workshop on Context-Awareness for Self-Managing Systems, pp. 38–34, CASEMANS '11 (2011)
5. Park, J., Sandhu, R.: The UCONABC usage control model. ACM Trans. Inf. Syst. Secur. **7**(1), 128–174 (2004)
6. Choi, J.H., Kang, D.H., Jang, H., Eom, Y.I.: Adaptive access control scheme utilizing context awareness in pervasive computing environments. In: 2008 IEEE International Performance, Computing and Communications Conference, 2008. IPCCC (2008)
7. Mayrhofer, R., Schmidtke, H.R., Sigg, S.: Security and trust in context-aware applications. Pers. Ubiquit. Comput. **18**(1), 115–116 (2014)
8. Temdee, P., Prasad, R.: Security for context-aware applications. In: Context-Aware Communication and Computing: Applications for Smart Environment. Springer Series in Wireless Technology. Springer, Cham, June 2017 (2018)
9. Duraivelu, V., Subramaniam, C.: A predictable information security based context aware trust model for organization management: a statistical analysis. J. Softw. **11**(6), 577–588 (2016)
10. Almutairi, S., Aldabbas, H., Abu-Samaha, A.: Review on the security related issues in context aware system. Int. J. Wireless Mobile Netw. **4**(3), 195 (2012)
11. Bardram, J.E., Kjær, R.E., Pedersen, M.Ø.: Context-aware user authentication—supporting proximity-based login in pervasive computing. In: International Conference on Ubiquitous Computing, pp. 107–123. Springer, Berlin, Heidelberg (2003, October)
12. OASIS. eXtensible Access Control Markup Language (XACML) Version 3.0, February 2009. Working Draft 8. [3]
13. W3C, "Platform for privacy preferences (p3p) project." [Online]. Available: http://www.w3org/P3P/
14. Shibboleth project. http://shibboleth.internet2.edu/
15. The Liberty Alliance (2007, October). http://www.projectliberty.org/ (accessed 15 Aug 2008)
16. Mahalle, P.M., Mahamure, S., Railkar, P.N., Chandre, P.R.: Distributed access control and authorization(DACA) for internet of things. IJETT **3**(3) (ISSN (Print): 2350—0808) (2016, September)

Chapter 8
Conclusions

Abstract This chapter gives a summary of all the chapters and concludes the book and presents the future outlook which can be initiated and investigated based on the theories and proposal presented. This book presents a detailed overview of context-aware pervasive systems and applications. The notion of a pervasive system, the perception of context and context management is presented and discussed in this book. State of the art for context management and context representation is also elucidated. Context-aware pervasive system and personalization are deliberated in the next part of this book. Context-aware profiling and context-aware clustering are presented and discussed in detail in the scope of this book. The next part of this book deals with different architectures for context-aware systems along with the role of machine learning techniques in context management. Context-aware computing in the telecom domain and security issues in the context-aware system is discussed in the last part of this book.

8.1 Summary

Context-aware system is a great technological revolution which represents the convergence of context management, information, communication, technology and machine learning. The envisaged context-aware applications present the vision of smart computing by making the devices and operations proactive. The seven chapters described in this book along with issues and challenges identified in the respective chapters are very important for context-aware systems and applications. Due to broadband technology and availability of the Internet widely, Internet computing has been evolved to the Internet of things and eventually to the Internet of everything. Modern scientific, engineering and business applications are increasingly dependent on a machine-to-machine communication, wherein there is less human intervention. This leads to the need for context-aware pervasive systems and applications where RFID, sensors and smartphones will play a key role to sense the context. Due to rapid development in wireless communication technology, the use of smartphone has increased drastically. With the advancement in full stack development, the use of mobile applications is also skyrocketing and eventually leads to several secu-

P. N. Mahalle and P. S. Dhotre, *Context-Aware Pervasive Systems and Applications*, Intelligent Systems Reference Library 169,
https://doi.org/10.1007/978-981-32-9952-8_8

rity challenges. Traditional methods of analysis desktop application are certainly not suitable due to resource constraints, mobility, distributed access and heterogeneity of contexts and devices. In the existing literature, the above-mentioned areas have been given little attention. The concepts and methods presented in this book have identified several important challenges for the design and development of context-aware applications. In the sequel, algorithms for security and clustering are also proposed and discussed in this book.

The book focuses on an overview of context and context management and how to perform context management in different use cases. In addition to this, the book also covers context-aware computing and personalization, various architectures for context-aware systems and security issues. This is intended to be a simple, accessible book and primary reference which puts forward best research roadmap, challenges and future outlook on context-aware pervasive systems and applications.

Chapter 1 presented an understanding of pervasive systems, emerging trends and economic significance. The role of sensors, actuators, RFID and Internet of things as technical building blocks in context-aware systems is also introduced in this chapter. Several issues and challenges in the context of design, technology and security are also discussed in this chapter. Finally, Chap. 1 concludes with an overview of the entire book.

Chapter 2 opened an overview of context, its classification and the notion of context-aware computing. In this chapter, state of the art on sensing the context using RFID, sensors and ontology is also presented and discussed. Chapter 2 finally concludes with a requirement for representing and exchanging the contexts in pervasive systems.

Chapter 3 describes the basic components and features of context-aware pervasive systems in general. A discussion on context-aware networks along with state of the art is also presented in the next part of this chapter. Issues and challenges with respect to routing, mobility and data management are also discussed in the last part of Chap. 3.

Chapter 4 elucidated how context-aware computing is important for personalization. The relationship between machine learning, ontology learning and personalization is also discussed in this chapter. The next part of this chapter focused on context-aware profiling and profile translation towards context-aware services. The need for clustering in ubiquitous computing and how context-aware clustering will play an important role in the next-generation Internet is discussed in the last part of this chapter.

Chapter 5 initially presents architectural requirements, different approaches and various platforms for context-aware applications. This chapter has also presented various context-aware algorithms along with design process and system issues. Last part of this chapter presented various case studies of context-aware systems and machine learning.

Chapter 6 described telecommunication context-aware services like content, session, location and identity. Telecom service-oriented architecture is presented and discussed in the next part of this chapter. The context-aware business model is also discussed in this chapter.

Chapter 7 discussed various security issues in context-aware systems. Importance of trust and security in pervasive systems is also presented in this chapter. Next part of this chapter has focused on very important issues of activity modelling and threat taxonomy in context-aware systems as well as corresponding mitigations.

8.2 Applications

Context-aware computing has a wide range of applications in every field like automobile, agriculture, transport, web computing and health care. Underlined applications of context-aware computing are based on the context classification like external/physical context which is basically sensed by sensors and includes parameters like location, light and sound. Internal/logical context includes parameters captured by users which include goal, emotion or presence. Location-based services include a wide varied range of applications which are listed below:

Location-aware services: Finding physical services like hotels, schools, malls, ATMs and hospitals and electronic services like access points, printers and hot spots.

Service improvement using location: Network services are improved by integrating context into applications. Incoming and outgoing calls also adapt to the location for service improvement.

Information Provider: Useful information is provided using by associating context with an underlined application which includes tourist guides, advertisements and recommendations.

Location awareness: Informing others about individual location helps a lot about the popularity of some event or group activity.

Security: Access control schemes and privacy solutions also use location which includes signing into Google applications and mobile applications.

Location-based services include other applications like entity tracking, shopping guides, travel and tourist guides, travel planner, geocaching, mobile games, location-aware call handling, indoor routing, car park assistance, traffic management, road tolling and fleet scheduling. Location-aware social networks (LASN) are the most popular and emerging application of context-aware computing. Location recommendation in LASN based on user's location history is also extensively used. LASN applications like Facebook, Instagram, Foursquare and Loopt are few examples where users can share photos, comments or check-in with respect to location. Location-aware services help to save time, money and enhance convenience and security.

Google map is the most useful and widely used location-aware service which uses the concept of web mapping which provides a satellite view of the cities, street maps and real-time traffic details. Google map continuously collects real-time data from various sources and post it on the cloud. Geographic information is acquired and stored using geospatial technology which includes GIS, GPS and remote sensing. Spatial databases are used to store this geographical information like maps and real-time data.

Apart from the applications mentioned above, there are many live use cases developed by enterprises and got a lot of popularity. Context-aware dynamic lighting of MIT living laboratories is very interesting to use case where office activities are mapped to lighting activities and corresponding colour settings. The main objective of this smart and dynamic lighting is to minimize energy consumption. Roof lights, fans and its gravity, speed, respectively, are also monitored and adjusted proactively depending on the requirement and number of persons in the room. The complete use case can be viewed on YouTube.[1]

Gimbal context-aware platform is most popular integrated platform developed by Qualcomm retail solutions. This platform keeps you updated about the metro schedules, lucrative restaurant offers, personalized services, recommendations based on the preferences for movies, games, etc. Gimbal context-aware platform transform the consumer experience by recommending personalized and timely recommendations to users on their mobile apps. Gimbals proximity beacons also enable location sensing to a more microlevel. Geofences is a useful feature in Gimbal which helps to turn location updates like entry and exit into offering by pushing notifications to the customers directly. The complete use case can be viewed at YouTube.[2]

The context-aware music player is useful to use case where appropriate music is played depending on the mood of the person coming back from work. The moment person is back to home from work, the context-sensing module gets active and it starts sensing various contextual parameters like body movement, gesture, time and location and based on these contextual parameters, the type of mood is decided. This mood type is then mapped to song type and appropriate song of this type is played.

8.3 Future Outlook

However, apart from the several applications of context-aware computing, there are many challenges to designing such applications. One of the important challenges is to process numeric and categorical contextual information. Numeric context includes time like 1979, 5.30 A.M., 1 A.M. to 4 A.M., and categorical context includes time as the morning, evening, weekday, etc., a location like home, office, restaurant and companion like a friend and neighbour. Specific techniques and algorithms are required to process numeric context information. Other challenges include how to enable explanations by context. We cannot receive significant improvement unless we don't explicitly explain the requirements to the end-users. Similar challenges include the use of context for information filtering, designing the new user interface

[1] https://www.youtube.com/watch?v=TUHFe_YoZCk.

[2] https://www.youtube.com/watch?v=P3LX84jdfnE.

for explanation and implementation of user-centric evaluations. User intent and better prediction is a new challenge which is coming up.

Future outlook includes designing of the new user interface to collect context and user-friendly interaction, the design of a dynamic user interface to explain context-aware information retrieval and recommendation systems, avoiding discussion and disputes about user privacy. Another interesting future work will be to address user privacy issues and problems while collecting and using user's context. Context suggestions are another important area which needs more attention.

MIX
Papier aus verantwortungsvollen Quellen
Paper from responsible sources
FSC® C105338

If you have any concerns about our products,
you can contact us on
ProductSafety@springernature.com

In case Publisher is established outside the EU,
the EU authorized representative is:
Springer Nature Customer Service Center GmbH
Europaplatz 3, 69115 Heidelberg, Germany

Printed by Libri Plureos GmbH
in Hamburg, Germany